STELLAR ASTRONOMY

Stellar Astronomy

HISTORICAL STUDIES

Michael Hoskin

SCIENCE HISTORY PUBLICATIONS

ISBN: 0 905193 04 0

Copyright © 1982 by Science History Publications Ltd

First published in this format 1982 by
SCIENCE HISTORY PUBLICATIONS LTD
Halfpenny Furze, Mill Lane, Chalfont St Giles,
Bucks HP8 4NR, England

Printed in Great Britain by
University Library, Cambridge

CONTENTS

Preface iv

Introduction: Principles and Methods 1

Section A: The Emerging Science of the Stars
1. An Overview 5
2. Novae and Variables from Tycho to Bullialdus 22
3. Hooke, Bradley and the Aberration of Light 29
4. Goodricke, Pigott and the Quest for Variable Stars 37
5. Herschel's Determination of the Solar Apex 56

Section B: A Century of Speculative Cosmologies
1. Introduction 67
2. Newton, Providence and the Universe of Stars 71
3. The Cosmology of Thomas Wright of Durham 101
4. The Cosmology of J. H. Lambert 117

Section C: The Riddle of the Nebulae
1. William Herschel's Early Investigations of Nebulae 125
2. The Nebulae from Herschel to Huggins 137
3. Island Universes: An Overview 154
4. Ritchey, Curtis and the Discovery of Novae in Spiral Nebulae ... 168
5. The 'Great Debate': What Really Happened 175

Acknowledgements 189

Index 191

Preface

The study of the planets was arguably the earliest branch of the natural sciences to reach maturity: planetary kinematics arrived at a high level of mathematical sophistication in Antiquity, and after 1687 the dynamical theory of the solar system embodied in Newton's *Principia* was to be the model for future scientific theories. By contrast, when Galileo was born the stars were still little more than a backcloth to the motions of the planets, and they were to appear 'fixed' relative to each other, without exception, until Newton was an old man. Stellar astronomy – the study of the stars, as opposed to that of the planets – is therefore a comparatively young science; and despite the minor role that planetary theory plays in modern astrophysics, it is into the history of planetary theory that historians of astronomy have put the bulk of their effort, to the neglect of the history of stellar astronomy. Such books as have appeared in English are mostly devoted to the twentieth century, and the student wishing to study stellar astronomy in the previous three centuries must mine much of his material from the pages of 'learned journals'.

In the hope of encouraging the study of the history of stellar astronomy, I have brought together in this volume studies published over the years, mostly in *Journal for the history of astronomy*, along with material appearing here for the first time. In Sections A and B, the first chapter provides the general context for the detailed studies that form the remaining chapters; in Section C the same role is played by chapters 2 and 3. In the interests of economy the *JHA* studies are reprinted with only minor changes, and I ask the indulgence of the reader when an occasional overlap results.

M.A.H.

INTRODUCTION

Principles and Methods

Light reaches us from the Sun in eight minutes; but even the nearest known star is more than four light-*years* away. Early stellar astronomers therefore found themselves studying objects immensely remote compared to the familiar planets of the solar system. Early modern cosmologists – who, as we shall see, invariably had a strong theological or philosophical commitment – were reasoning about the most distant objects in the universe, whether visible or not. It is therefore to be expected that in stellar astronomy in general, and in cosmology in particular, the theorists whose work we shall study will be operating at and beyond the limits of the evidence and that methodological principles will therefore play an exceptionally important role in their thinking. At the same time, novel concepts had to be introduced to meet new problems; and novel techniques devised to extract theoretical consequences from the limited data available. By way of introduction, therefore, we alert the reader to five of the principles, concepts and methods that play a significant role in the pages that follow.

Simplicity

As to the steady-state theorist of the mid-twentieth century, so to the cosmologist of the eighteenth or nineteenth century the simplicity of the universe is of fundamental importance in his theorising. Since he is by definition theorising about the most remote objects accessible with contemporary apparatus (and even beyond), the evidence at his disposal is inevitably scanty and ambiguous, and observations alone are likely to be ineffectual in deciding between rival theories. We shall several times encounter an almost-ruthless insistence on a simple theory, and a marked reluctance to reconcile disparate observations by the easy option of a multi-faceted explanation. For example, in chap. 1 of Section C, we shall find William Herschel insisting that the milky patches known as nebulae were vast star systems disguised by distance, meanwhile closing his mind to the changes he believed he had himself observed in the Orion Nebula – changes which could not have occurred so quickly if the nebula was indeed a vast star system. A century later, William Huggins (see Section C, chaps. 2 and 3) used spectroscopy to establish beyond question that *some* nebulae are gaseous; but when the Andromeda Nebula was found to have the spectrum to be expected of a galaxy of stars, Huggins devised a physical theory for the nebula that would enable it nevertheless to be classified with the gaseous nebulae. Neither man was happy with the easy but multiple (and correct!) solution, that nebulae come in many different kinds; each preferred a simple and therefore powerful theory even if this appeared to be in conflict with the evidence.

Uniformity

Closely related to simplicity is the more specific assumption of uniformity: that celestial bodies may be presumed to be of similar nature, to lie at similar distances, to move at similar speeds, and so forth, until proved to the contrary. It will often be the case that an answer, however provisional, can be made to a question only if gaps in the evidence can be filled by an assumption of uniformity. This is notably the case with measurements of very great distances. James Gregory, in *Geometriae pars universalis* (Padua, 1668; p. 147), had spoken of the Sun as *stella fixa vicina,* "the local fixed star". Newton, as we shall see (Section B, chap. 2), established the distances to the nearest stars by adapting a technique of Gregory's which depended upon the stars being highly uniform with the Sun and their faintness therefore entirely the consequence of their great distances; for an illustration of the same concept and methods at work in modern astronomy, one only has to read the chapter on "The measurement of astronomical distances" in F. Hoyle's *Frontiers of astronomy* (London, 1955). In the absence of direct trigonometrical measurements of annual parallax, Newton's assumption permitted provisional answers to be given when otherwise the question would have gone unanswered. Similarly, William Herschel (Section A, chap. 1) derived an outline for our own Galaxy by assuming that within the Galaxy the stars are uniformly dense, so that the more stars there are in a given direction, the further the Galaxy extends in that direction. Later his own studies of star clusters forced him to recognise that the distribution of the stars is very far from uniformly dense, so much so that his outline of the Galaxy must be withdrawn; but students of nature abhor a vacuum, and it is a striking indication of the pressure to supply an answer, any answer, to a scientific question, that his outline, though disowned by its creator, was still being reproduced decades after his death.

The Gregorian assumption that the stars are uniform in physical properties was especially necessary to Herschel in that it offered the only hope of establishing the distribution of the stars in three dimensions, the 'construction of the heavens' that was his principal goal in astronomy. Yet even in Newton's day there were warning signs to be seen, for example in the new and variable stars (Section A, chap. 2). Herschel personally observed binary stars (Section A, chap. 1), pairs of companion stars which in some examples differed strikingly from each other; and during his lifetime certain faint stars, notably 61 Cygni, were discovered to have rapid 'proper motions' across the sky, which implied that they must be near notwithstanding their faintness. Herschel's career, which occupies the central ground in the studies of this volume, can be seen as a desperate rearguard action in defence of the assumption of uniformity among the stars, and the abandonment of the assumption by the next generation accounts in part for the poverty of their cosmological thinking.

Time

For Newton and his contemporaries, God was the great clockmaker and the universe a machine in which changes were cyclic, as in the motions of the planets around the Sun. The last great exponent of such a universe was J. H. Lambert (Section B, chap. 4). The stars presented a special problem (at least until 1718 when Halley announced the first proper motions); for gravity might be expected to wreak havoc among a system of stars initially 'fixed' and at rest. As William Whiston put it in his *Astronomical principles of religion* (London, 1717, pp. 88-89), "it follows, that the several Systems, with their several Fixed Stars or Suns, do naturally and constantly, unless a Miraculous Power interposes to hinder it, approach nearer and nearer to the common Center of all their Gravity"; in other words, gravitational collapse. How Newton sought to escape from this difficulty is the subject of Section B, chap. 2, but the possibility of fundamental change with time – what we might loosely call the 'evolution' of the universe – had been faced.

That the velocity of light is finite had been shown by seventeenth-century studies of the timings of the eclipses of Jupiter's satellites; but the fact became of direct concern to every astronomer committed to making precise observations when Bradley showed (Section A, chap. 3) that such observations are modified by the direction of the Earth's velocity at the time of observation. It was recognised, of course, that the observer sees a star, not as it is at the moment of observation, but as it was when the light began its journey from the star. The difference became significant, indeed awe-inspiring, when one looked as far into space as Herschel. As the poet Thomas Campbell reported to a friend in a letter written in September 1813, Herschel told him: "I have observed stars of which the light, it can be proved, must take two million years to reach the earth ... if those distant bodies had ceased to exist millions of years ago, we should still see them, as the light did travel after the body was gone" (*The Herschel chronicle*, ed. C. A. Lubbock (Cambridge, 1933), 336).

It was Herschel, too, who recognised how the astronomer can, so to speak, observe changes that in reality take enormous lengths of time (see Section C, chap. 2). He parades for our inspection similar celestial objects at different stages of their development, asking: "Is it not almost the same thing, whether we live successively to witness the germination, blooming, foliage, fecundity, fading, withering, and corruption of a plant, or whether a vast number of specimens, selected from every stage through which the plant passes in the course of its existence, be brought at once to our view?" (*Phil. trans.*, lxxix (1789), 226).

Stellar Statistics

This basic tool of modern astronomy was first employed in mature form by Isaac Newton, when he analysed the implications of the numbers of the stars of successive magnitudes (see Section B, chap. 2) in drafts of a theorem only the barest hints of which saw publication in his own day. William Herschel in one of his last papers (*Phil. Trans.*, cvii (1817), 302-31) unknowingly recapitulated Newton's analysis to an astonishing extent, but the technique of the analysis of stellar statistics came into the possession of astronomers when it was employed in mature form by Herschel earlier in his career, in the investigation of the outline of our Galaxy already mentioned. Herschel counted the numbers of stars in ten neighbouring fields of view, averaged these numbers to obtain a 'Star-Gage' [*sic*], and interpreted the result as proportional to the cone-shaped volume of space within the Galaxy seen by the observer in that direction; the 'height' of the cone was then proportional to the distance to the border of the Galaxy in that direction.

Probability Theory

Lastly, we draw attention to John Michell's brilliant paper in *Philosophical transactions* (lvii (1767), 234-64), in which (see Section A, chap. 1) he argues that the number of multiple stars is so great that they must be true stellar associations rather than chance, optical effects, because of "*the greatness of the odds against things having been in the present situation, if it was not owing to some such cause*" (p. 243, italics supplied). Basing himself solely on this probability argument, Michell was able to offer evidence that gravity (or a similar force) operates beyond the solar system and among the stars; and to warn that most double stars would be unsuitable subjects for distance measurements which depended on the two stars being remote from each other.

A. THE EMERGING SCIENCE OF THE STARS

1. *An Overview*

In this chapter we set the scene for the four detailed studies that follow, by outlining the emerging science of the stars as individuals, from the seventeenth century to the rise of astrophysics in the nineteenth. The study of nebulae and systems of stars and of the universe as a whole belongs to the later sections of this book. Here we consider the individual stars, and especially the stars (and the Milky Way) visible to the naked eye.

The distances of the stars

Copernicus had failed to observe what his theory predicted, the apparent annual movements among the stars which reflect the annual movements of us, the observers, as we orbit the Sun; that is, he had failed to detect 'annual parallax'. It is often remarked that annual parallax was first detected only in the 1830s. Yet this is misleading: first, because from the time of Robert Hooke there were frequent claims from observers who were convinced that they *had* successfully measured annual parallax, and thereby determined the distances of stars, and so to leap the three centuries from Copernicus to Bessel is to conceal the confusion created by these mistaken claims; second, and more seriously, because *the correct scale of the distances of the stars was recognised by Newton,* confirmed by Bradley, and accepted by all subsequent generations.

This remarkable achievement required an hypothesis: that all the stars, including the Sun, give out roughly the same amount of light. This being so, and provided there is no significant loss of light as it travels through space, then the apparent brightness of a star is a measure of its distance from the observer, the apparent brightness diminishing with the square of the distance. For example, if the Sun appears n^2 times brighter than Sirius, then Sirius is n times further than the Sun (that is, its distance is n astronomical units). The problem then reduces to the practical one of actually comparing the brightness of the Sun with that of Sirius.

This problem was brilliantly solved in 1668 by James Gregory. Gregory's technique was to make Jupiter a bridge (so to speak) between the Sun and Sirius. We are to consider Jupiter at a time when it appears equal in brightness to Sirius. The apparent brightness of Jupiter (and therefore of Sirius) equals the brightness of the Sun multiplied by a factor which is a function of dimensions within the solar system (taken as known) and of the ability of Jupiter to reflect light. Gregory's calculations put Sirius at 83,190 astronomical units, but he recognises that he has used obsolete values and that this is an underestimate.[1]

Christiaan Huygens in his *Cosmotheoros*,[2] published in 1698 and widely read, used a different and somewhat clumsy technique. He observed the Sun through a tiny hole in a disk, hoping that the visible portion of the Sun would appear as bright as Sirius; he intended to compare the light from these two stars by calculating what fraction of the Sun could be seen through the hole. But it proved impossible to make a hole small enough, and Huygens had to reduce the light still further by means of a lens. In this way he arrived at a distance of 27,664 astronomical units, a figure that appears repeatedly in the literature of the first quarter of the eighteenth century. By contrast, Gregory's technique seems to have been noticed only by Newton (see Section B, chap. 2 below), and by mischance Newton's astonishing figure of about one million astronomical units was unpublished until 1728, more than forty years after it was first derived. As the distance of Sirius is somewhat over half-a-million astronomical units, Newton's result is excellent, though this is in part a matter of luck since the underlying hypothesis of uniformity among the stars is hopelessly ill-founded.

Gregory's method, once known, quickly superseded that of Huygens; it was to be used into the nineteenth century,[3] and each new application confirmed that the light from the nearest stars takes some years to reach us – a remarkable insight for the infant discipline of stellar astronomy.

Such a uniformity assumption may appear arbitrary and unjustified: Gregory and Newton had no *evidence* that the stars all give out the same amount of light. The justification lies in the possibilities opened up by such an assumption; and such assumptions have become a regular component of distance measurements in astronomy, since the direct application of trigonometry to distance measurement is inherently limited by instrumental errors and therefore can be used only with the nearer celestial objects. At greater distances other methods must be invoked, and these will often involve assumptions of uniformity of Cepheid variable stars, of globular clusters, of galaxies, or whatever. The problem is to know which celestial objects form, so to speak, a true species, the members of which differ from each other only to a limited extent.

In this sense, the stars as a whole do not form a species, and it was William Herschel's misfortune to commit himself to a programme – the investigation of 'the construction of the heavens' in three dimensions[4] – for which measurement of great distances was fundamental, at the very time when evidence of the possible disparity between one star and another was mounting. He himself was the first to observe pairs of stars (or 'binary' stars) in which each member of the pair moves in orbit around its companion, so that the two stars lie at the same distance from us.[5] In some instances Herschel specifies that the two stars are of unequal apparent brightness (in which case, as they are at the same distance from the observer, the inequality is not only apparent but real). Yet when he subsequently needed to compare the distances of two stars, he continued to assume that relative apparent brightnesses could be converted into relative distances with an accuracy of three significant figures![6]

Of course, even if the uniformity assumption was granted, there remained the practical difficulty of actually comparing the apparent brightness of two stars. Perhaps surprisingly, Gregory's success in the extreme case where one star was the enormously bright Sun was not to be extended to the more delicate comparison of two faint stars for another century and a half. Meanwhile, both Newton (in manuscript[7]) and Herschel (in print[8]) assumed that the traditional magnitudes inherited from Antiquity could be taken as a direct index of distances, so that a star just visible to the unaided eye, and therefore of the sixth magnitude, would be six times more distant than a bright star of the first magnitude. Both Newton[9] and Herschel[10] eventually found that the resulting distribution of the stars in three dimensions was implausible, and that sixth magnitude stars must be placed at somewhat greater distances. Towards the end of his career, Herschel developed the technique of using two similar telescopes and pointing one at each of the two stars whose apparent brightness he wished to compare. The telescope pointing to the brighter star was then reduced in aperture until the two stars appeared equally bright, and the ratio of the two apertures provided the ratio of the brightnesses.[11] Later, John Herschel at the Cape of Good Hope in the 1830s devised a method for comparing the brightness of a star with that of an artificial star formed by the total reflection of moonlight from the base of a prism,[12] and this inaugurated the advances in photometry which led in the last quarter of the century to the great Harvard, Oxford and Potsdam catalogues of stellar magnitudes.[13]

The "conjectural" method had served a valuable purpose in permitting theorising which otherwise could not have been attempted. But many astronomers rejected the underlying assumption; and even if the assumption had been approximately true, important problems depended for their solution upon *precise* knowledge of stellar distances.[14] Given the distance of a star, its apparent brightness would give its *intrinsic brightness* (and, incidentally, test the hypothesis of uniformity); its observed proper (that is, individual) motion across the sky would give its *velocity at right angles to the line of sight,* and such velocities will (collectively) reflect the *velocity of the solar system* itself; and, if Newton's laws applied among the stars, the *combined masses* of two stars in orbit around each other could be obtained from observations of their angular separation and period of rotation. The prizes were tempting; but the difficulties seemed to increase rather than decrease with the passage of time.

The first modern-style attempt to measure annual parallax (see Section A, chap. 3) was that of Robert Hooke in 1669 – modern style, because he is acutely aware of the practical difficulties involved in making measurements of an annual, cyclic phenomenon with instruments sensitive to seasonal changes of temperature and humidity.[15] Like subsequent investigators, Hooke selected for study a star (in his case, γ Draconis) which was thought to be *near* (because it was bright) and which was *favourably situated* (because it passed overhead in London and observations of it were therefore free of the complication of atmospheric refraction). His pitifully few observations suggested that the (double) angle he was seeking to measure was about 30″. Much more convincing was the series of observations of the Pole Star which Flamsteed carried out over a period of seven years;[16] but Flamsteed had made a geometrical error and in fact the displacements he was measuring were at right angles to those predicted by theory. It was James Bradley (see Section A, chap. 3) who continued Hooke's investigation of γ Draconis and who realised that the observed displacements were due to the 'aberration' of light; that is, they were caused by the *velocity* of the observer (in a direction *tangential* to the Earth's orbit), rather than by his *position* away from the centre of the solar system (in a radial direction).[17] Bradley's discovery, announced in 1729, was physical proof at last of the velocity of the Earth and so of the Copernican hypothesis, and, with his later discovery of the nutation of the Earth's axis,[18] ushered in the modern era of exact positional astronomy: all previous positional astronomy had been marred by these two unsuspected effects.

Bradley's meticulous measurements made it clear that the stellar distances derived by Hooke and Flamsteed (and Huygens) were illusory. But he himself failed to detect an annual parallax for γ Draconis. From his failure, and his knowledge of the accuracy of his instruments, he inferred that γ Draconis must be at least 400,000 astronomical units away[19] – a result which accorded well with the results of the "conjectural" method as developed by Newton forty years earlier but published (posthumously) only in 1728.[20] The appearance of these two works within a few months of each other, each implying (and for quite different reasons) that the nearest stars are at distances to be measured in hundreds of thousands of astronomical units, convinced astronomers of the scale of interstellar distances; and this understanding dates from the time of Bradley and not, as often said, from the 1830s.

Not surprisingly, Bradley's revelation of the almost incredible delicacy of the required measurements (one second of arc for the double parallax, or the width of a coin at several kilometres' distance), and the apparent near-impossibility of maintaining such accuracy over an annual cycle, resulted in a failure of nerve among those few astronomers who possessed instruments capable of precision measurements. But in 1802, Giuseppe Piazzi at Palermo noticed anomalies in his observations of α Lyrae which he thought might be due to parallax.[21] α Lyrae (Vega) is the fourth brightest star in the sky, and so presumably near, and Piazzi decided to observe several more of the brightest stars – Capella, Aldebaran, Sirius, Procyon, Arcturus, Altair and Vega – at times when they would show the maximum parallax. He measured their distances from the zenith with a Ramsden circle which he rotated through two right angles between one night and the next so as to control and compensate for certain instrumental errors. His results, published in 1805, he considered doubtful, though he believed they showed some effect of parallax.

Giuseppe Calandrelli found that his observations of Vega, taken in Rome at different times of year from those of Piazzi, gave significantly different results, and he therefore made observations of this star in 1805 and 1806.[22] But it seems that the instruments of both Piazzi and Calandrelli were subject to temperature effects, and their parallaxes of several seconds never commanded the assent of their professional colleagues.[23]

A sustained attack on the problem of parallaxes was carried out between 1808 and 1822 by John Brinkley of Dublin, with a fine eight-foot circle begun by Ramsden and completed by Berge. This circle could be rotated, like that of Piazzi, and this similarly permitted the observer to eliminate errors. Brinkley measured the zenith distances of several of the brightest stars – Arcturus, Vega, Altair, Deneb – and of the Pole Star, of γ Draconis (which passed overhead), and of several other stars besides. His series of papers[24] also includes careful discussion of the innumerable interfering factors. Brinkley deserved well of astronomy for his sustained and critical campaign, but his parallaxes of several seconds were illusory. By a cruel stroke of fortune, much less careful measurements of the polar distances of Vega, Altair and Deneb between 1813 and 1823 by Pond at Greenwich with the mural circle of Troughton led Pond to the correct conclusion that the parallaxes were too small for either of them to measure. Pond undertook this investigation as a test of Brinkley's results, a test which Brinkley at first welcomed, but relations between the two men degenerated to produce one of the most notorious scientific controversies of the period.[25]

Piazzi, Calandrelli, Brinkley and Pond had each selected for examination *bright stars,* supposing these to be the nearest, and had measured their *angular distance from the pole or the zenith,* a large angle whose measurement was likely to be affected by any one of a dozen sources of error. Gradually, however, astronomers were realising that *brightness is not the only criterion of nearness.* Halley had noticed[26] in 1718 that certain stars were moving across the sky with proper (or individual) motions, and astronomers had since recognised in a large proper motion an indication of nearness: if we are in a crowd, it is those closest to us who appear to us to be moving most quickly.[27] For a long time there was little clash between the criterion of brightness and the criterion of large proper motion, for in most cases the stars examined for possible proper motion were bright ones; but in 1812 F. W. Bessel demonstrated at length that a faint (fifth magnitude) star in the constellation of the Swan, 61 Cygni, had the surprisingly large proper motion of over five seconds per annum.[28] That same year three astronomers examined it for parallax: François Arago and C. L. Mathieu measured its distance from the zenith, and inferred a parallax of over half a second, but the result was not published for over twenty years;[29] however, Baron de Lindenau, between 1812 and 1814, measured its distance, not from the zenith or the pole, but from nearby stars,[30] and he thereby revived a technique whose history goes back to Galileo – the *double star method of measuring annual parallax.*[31]

Of the innumerable interfering factors which had bedevilled such parallax measurements over the years, most would affect (almost) equally the observations of two stars extremely close together in the night sky. To take only a single example, atmospheric refraction will affect equally the observed positions of the two stars, and its effects can therefore be circumvented if we measure the *differences* in angle between the two positions. Bradley was only one of several astronomers to recommend this technique to his successors,[32] and William Herschel had systematically searched the sky for suitable pairs of neighbouring stars, or 'double stars'.[33] Herschel was looking for double stars whose members were so close together in the sky that at first glance they appear as a single star. If the tiny angle between them was measured for size and direction with a micrometer attached to the telescope, even a slight change in the relative positions of the two stars should be noticeable.

A double star may occur either because two stars distant from each other happen by chance to lie in almost the same direction from Earth (an 'optical' double), or because two stars are physical companions of each other (and form a 'binary star'). A binary star would be useless for parallax measurements, because the two components, lying at the same distance from the observer, would show the same annual alteration of position. But in an optical double, the nearer star would appear to move proportionately more, and if the further star was very distant then for practical purposes it would behave as a fixed reference point.

Lindenau exploited this technique, but not with the conviction of a Herschel. He compared 61 Cygni, the "flying star", with bright stars some distance away across the sky, but without success. Bessel himself, in 1815 and 1816, compared the position of the flying star with others in the same constellation, but was defeated by an unknown instrumental error.[34]

Meanwhile, Wilhelm Struve with the great meridian refractor at Dorpat between 1818 and 1821 attempted observations of each of a pair of stars chosen because they lay in almost opposite directions from the pole, so that one of the pair had its superior culmination at about the same time as the other had its inferior culmination; various interfering factors could be avoided by combining the observations. Struve published results for some sixteen pairs,[35] but temperature changes seem to have vitiated his values.

It was the darkest hour before the dawn. Advances on two fronts were about to produce the decisive breakthrough: first, the strategy involved in the selection of the object and method of observation was to be reassessed; and second, instruments of far superior quality were to be employed.

First, *the object and method of observation.* The star to be studied must give convincing evidence of its nearness, and in a cogent review[36] of the criteria of nearness published in 1837 Struve set out three criteria: first, a star is likely to be near if it is bright; second, it is likely to be near if it has a large proper motion; third, a binary star is likely to be near if the two stars of which it is composed appear widely separated for the time they take to complete one orbit. Struve listed the stars which best satisfied each criterion and drew attention to those which satisfied more than one criterion. The value of his list is shown by the number of the nearest stars known today which it includes. As to the method of observation, both Struve and Bessel determined to use only pairs of very close stars, and to employ for the comparison a micrometer to give precise measurements.

Second, both men were privileged to have at their disposal *instruments far superior to those previously available.* At Dorpat, Struve now possessed a magnificent 24cm equatorial by Fraunhofer, for many years the largest refractor in the world. In 1835, he decided to attempt the measurement of the parallax of Vega, by reference to a faint (10.5 mag) star only 43″ distance away. He was confident that the two stars did not form a binary system because the faint star did not share the large proper motion of Vega. In 1837, Struve announced the results of seventeen observations he had made between November 1835 and December 1836.[37] He found a parallax of 1/8″ with a probable error of under 1/20″, a result that is almost precisely the modern value; but he promised a continuation of his observations, and in 1840 gave the results of 96 measurements up to August 1838.[38] Now the value obtained was more than double its predecessor, and *both* values were therefore subject to some doubt.

Meanwhile, at Königsberg, Bessel had applied the fine 16cm Fraunhofer heliometer to the flying star.[39] "I undertook to make this investigation upon the star 61 Cygni", he later wrote, "which, by reason of its great proper motion, is perhaps the best of all; which affords the advantage of being a double star, and on that account may be observed with greater accuracy; and which is so near the pole that, with the exception of a small part of the year, it can always be observed at night at a sufficient distance from the horizon".[40] In September 1834 he began to compare the flying star with small neighbouring stars which did not share the proper motion and so must be physically independent, but his observations were interrupted by other work, including the arrival of Halley's comet, and it was not until 1837 that he returned to the task, encouraged by the preliminary results of Struve. For more than a year he kept up his observations, commonly repeating them sixteen times every night and even more frequently when the seeing was especially good. Before the end of 1838, he announced that his results showed – correctly as we know – that the parallax of the star was about 1/3". But what lent immediate conviction that at long last the sounding line had touched bottom was the pattern of Bessel's observations, which followed so closely that predicted by theory.[41] The late Otto Struve has pointed out[42] that his ancestor's preliminary result was well-founded and prior to that of Bessel; but after so many disappointments, astronomers demanded observations of the frequency and consistency that only Bessel supplied.

It was only a few weeks later that Thomas Henderson announced the parallax of the southern star, α Centauri.[43] Not only was the star bright, but it had a large proper motion, and was a double star with a wide angular separation. It therefore satisfied all three of Struve's criteria for nearness; but it seems that the vital clue, the large proper motion of the two components, was unknown to Henderson until after ill-health had led to his return from the Cape of Good Hope in 1833. It was only when subsequently reducing his observations of the declinations of the stars that he discovered the large proper motion and therefore looked for evidence of parallax; even when parallax showed up in his figures, he hesitated to announce the result until a similar parallax was found in the observations of right ascension taken by his colleague. So far as is known today, α Centauri is the Sun's nearest neighbour among the stars and therefore has the largest parallax of all, about 3/4". But it is doubtful whether Henderson's results, derived as they were from observations made with inferior instruments and for another purpose, would have carried conviction had not the stage been set by Bessel's announcement.

At long last, distances of stars had been directly established and a yardstick established for the calibration of future measurements of far greater distances. Success had come, first from the great advance in instrumentation, for which the credit belongs to Fraunhofer; second, from the careful and intelligent consideration of the criteria of nearness, and especially the realisation of the significance of proper motion in determining both nearness and the presence or absence of physical association; and third, from the frequency and care with which Bessel (and Struve) carried out the observations and so revealed a convincing consistency of pattern in the results. Their successors did not always follow their example: most of the nearest stars are faint like 61 Cygni and many are not even visible to the naked eye, but astronomers were lured into concentrating their efforts on the brighter stars, to the neglect of the other criteria of nearness, and it was another half-century before significant further progress was made.[44]

The positions and magnitudes of the stars: solar motion
Greenwich Observatory was founded in part, and the resulting 3000-star "British catalogue" of Flamsteed published, in order to establish an exact frame

of reference for the observation of the movements of the Moon.[45] But gradually the value of such catalogues for the theoretical development of stellar astronomy was recognized. Given precise positions for a star at two dates, and we know the proper motion of the star across the sky. If two stars which appear close together in the sky share the same proper motion, they probably form a binary system and we shall wish to study the behaviour of this system; if they do not share the same proper motion, they form an optical double and so may prove a suitable subject for observations to determine annual parallax. And from an overall study of the known proper motions of stars, we may uncover a pattern in them which in fact reflects the motion of us, the observers, and of the solar system to which we belong.

The accuracy with which the proper motion of a given star is known depends on the length of time separating the two observed positions and the accuracy with which each of these two positions has been observed. The early nineteenth century might contrive ways of improving *new* observations; but how could the accuracy of observations made by past generations be improved! Clearly, sufficiently exact positional astronomy had been impossible before Bradley revealed the need to make allowance for aberration and nutation. Fortunately, he had himself subsequently carried out very exact observations of the stars from 1750 to 1762, with excellent instruments and meticulous attention to sources of error; but for many years these data had remained in manuscript, and even when published[46] in 1798-1805 they were unreduced – that is, they were still without the necessary corrections for aberration and other interfering factors. It was Bessel who possessed the mathematical virtuosity and patience to analyse the errors of Bradley's instruments, to effect the necessary reduction, and so to extract from the precious data bequeathed by Bradley, accurate positions of 3,222 stars as in the year 1755. His volume, published in 1818 with the proud title of *Fundamenta astronomiae pro anno 1755,* carried the beginnings of precision astronomy back to 1755.[47] The lasting significance of Bradley's observations is shown by the appearance of a *New reduction* in three volumes published by G. F. J. A. von Auwers in 1882-1903.[48]

Meanwhile, there appeared new catalogues of positions and brightnesses of stars (including multiple stars) of improved accuracy and ever-increasing thoroughness. For example, observations made under Argelander's supervision from 1852 to 1859 resulted in the three volumes of the *Bonner Durchmusterung* (1859-62) with accompanying atlases, giving nearly 325,000 stars and providing a sky survey of unrivalled scope. It was David Gill who first realised the value of photography in such a task, when in 1882 he found numerous stars on a photograph he had taken of a comet.[49] By the end of the century his *Cape photographic Durchmusterung*[50] gave the places of nearly half a million stars, and he helped to launch the *Carte du ciel,* a photographic chart showing stars to the fourteenth magnitude with a catalogue giving precise positions of two million stars, which was not completed until recently.[51]

The detection of slight *changes in brightness* of stars was not possible until William Herschel hit on the idea of arranging stars in order of decreasing brightness, so that any future change would disturb the delicate ordering and so be recognisable.[52] But gross changes had been known (see Section A, chap. 2) since the appearance of "Tycho's nova" of 1572, which had vanished after sixteen months. In 1638, a star in the Whale later to be known as Mira Ceti appeared and then vanished, only to reappear once more after a few months; and soon there were frequent reports of such variable stars. In fact Mira is invisible to the naked eye for nearly half the time, but at its brightest it can reach second magnitude, and so it was a likely candidate for early detection. In 1667 Ismael Bullialdus published a study[53] of the observations made of Mira since 1638, and showed that it varied with a period of 333 days: although its brightness at maximum could not be predicted, the date on which the maximum would occur could, and to this extent one variable star was now known to be law-like in its behaviour. Bullialdus suggested that the star was partly dark, and was rotating with this period;[54] but, as Whiston was to emphasise, superimposed on the regular period were marked irregularities of brightness.[55]

In an age which had newly realised that the Sun is a star, a facile explanation of Mira and the growing number of other variable stars was ready to hand. Sunspots are carried round with each rotation of the Sun, and so the Sun's brightness must regularly fluctuate; but the sunspots themselves gradually appear and disappear, so the Sun's brightness must also fluctuate irregularly. In other words, all changes in variable stars, whether regular or irregular, could be assimilated to sunspots.[56] True, the sudden appearance of a bright new star was difficult to explain on these lines; but perhaps such a dramatic increase in brightness resulted from the collision of a comet with the star,[57] or from the tilting of the axis of the star (envisaged as disk-shaped) as a consequence of the near passage of a planet or comet.[58]

So, with the exception of novae, the sunspot analogy explained everything – and nothing. Even John Goodricke, in suggesting (see Section A, chap. 4) in 1783 that the variation in the star Algol, with a regular cycle of some 69 hours, might be explained as occultations by a dark companion star, included in the very same sentence an alternative explanation in terms of sunspots.[59] And so, until the advent of the spectroscope revolutionised the evidence available, the explanation of novae and variables remained at a virtual standstill.[60]

The first announcement of *proper motions* of stars came in 1718, when Edmond Halley pointed out that three bright stars were no longer in the positions recorded for them in Antiquity.[61] Until then, the 'fixed stars' had fully justified their ancient title, and it had seemed to Newton (Section B, chap. 2) that the stars were evidently motionless in space, despite the action of gravitational attraction. His contemporaries Hooke and William Whiston had been less dogmatic. Hooke, writing before 1669, thought the discrepancies between stellar positions recorded by observers of the calibre of Tycho and Hevelius probably genuine; Whiston, writing on the eve of Halley's announcement, accepted that no authentic changes of position were known, but did not regard this as proof that the Sun is at rest: but it must be moving slowly – "unless it be mov'd uniformly and evenly with the Centers of other Systems".[62] As to the causes of proper motions, in 1748 James Bradley (like Hooke before him) remarked that observed motions were relative and could arise either from movements of the stars themselves or from the motion of the solar system.[63]

The pattern of proper motions which would reflect the actual motion of the solar system was described in 1760 by Tobias Mayer of Göttingen.[64] As when a man walks through a forest, the trees in front of him appear to open out as he approaches while those behind him appear to close up, so the proper motions of stars will diverge from the point of the sky towards which the Sun is travelling and converge towards the point which the Sun is leaving.

Mayer himself could see no such pattern in the proper motions known to him, and concluded that the Sun is at rest. It was William Herschel, in 1783, who first found such a pattern.[65] In a daring but questionable analysis of data published by other astronomers (see Section A, chap. 5), he argued that most of the known proper motions simply reflect the Sun's motion towards the star λ Herculis, and since Fortune favours the brave, his result is close to the modern position. Shortly afterwards Prévost and Klügel announced directions in fair agreement with that of Herschel.[66]

Never one to let well alone, Herschel returned to the subject in 1805 and 1806, and attempted to establish not only the direction but also the speed of the solar motion.[67] To convert observed proper (that is, angular) motions into actual speeds at right angles to the lines of sight, Herschel required a knowledge of distances, and for this he had to invoke the assumption that all stars are equally bright. As this assumption is, in specific cases, hopelessly ill-founded, Herschel was led into a tangle of argumentation.[68] In particular, he concluded that bright (and therefore near) stars with no known proper motions must in fact be moving uniformly with the Sun, so that Herschel ended his analysis with more motions than he started with.

The number and accuracy of known proper motions grew sharply with the publication of Bessel's *Fundamenta,* but Bessel himself could find no pattern to reveal the direction of solar motion.[69] Such was his prestige that his opinion was generally accepted until 1837, when Argelander published a very extensive investigation which essentially confirmed Herschel's 1783 direction and permanently established the reality of the movement of the solar system through space.[70]

Studies of the motion of the solar system led to renewed speculation concerning the position round which the Sun and the other stars were presumed by some to orbit under gravitational attraction. Yet Newton himself had offered no positive evidence that attraction extends beyond the solar system. The first such argument based on observations came in the magnificent paper by John Michell in the *Philosophical transactions* for 1767.[71] There he asks how many double stars and clusters of stars would be expected if the stars were simply scattered at random throughout space. "If now we compute ... what the probability is, that no two stars, in the whole heavens, should have been within so small a distance from each other, as the two stars β Capricorni ... we shall find it to be about 80 to 1." As to the six brightest stars of the Pleiades, "we shall find the odds to be near 500000 to 1, that no six stars, out of that number, scattered at random, in the whole heavens, would be within so small a distance from each other, as the Pleiades are".[72]

With this epoch-making probability argument, Michell shows that it is highly likely that the members of a particular double star or cluster are companions held together by physical forces, rather than being isolated stars which happen to lie in the same direction from the observer; in other words, most double stars are binaries. When William Herschel began his search for (optical) doubles to facilitate the measurement of annual parallax, he was unaware of Michell's paper, and Michell published a further paper forecasting that many of Herschel's doubles would prove to be binaries.[73] When Herschel re-examined some of his doubles in 1803, he found that in several instances Michell's reasoning was confirmed and the components had indeed moved in orbit around each other[74] – though Herschel could not prove that the force at work was actually gravitational attraction; this had to await the careful geometrical analysis of Savary, Encke and John Herschel a quarter of a century later.[75]

It was not long before these investigations yielded dramatic results. Once the distance was determined for a binary star of known orbit, the space separating the two components could be established, and Newtonian mechanics then supplied the *combined masses* of the stars. Again, Bessel had noticed that both Sirius and Procyon were moving with proper motions that varied – possibly, he suggested to John Herschel, because each was also in orbit about an unseen companion star.[76] In September 1861 the position of the hypothetical companion of Sirius was calculated,[77] and the following January the star was actually observed by Alvan G. Clark;[78] Procyon's companion was seen in 1896.[79] Both are *white dwarfs* and even in the nineteenth century enough of their remarkable properties were recognized to create astonishment.[80]

The Milky Way

When we consider the large-scale distribution of the stars, the most remarkable fact of observation is the Milky Way. Galileo's telescope had confirmed the common opinion of his day,[81] that the Milky Way is the optical effect of very many faint stars, and in 1750 Thomas Wright of Durham suggested (see Section B, chap. 3) that the Sun is one of many stars which orbit around the Divine Centre of our star system.[82] Wright held that the system of the stars is spherical, or else is a flattened but hollow ring; in the latter case the stars were "all moving the same way, and not much deviating from the same plane, as the planets in their heliocentric motion do round the solar body".[83] His opinion was partially misunderstood by Immanuel Kant, who gratefully seized on this analogy between the Milky Way and the plane of the solar system and derived from Wright a disk-theory of the Galaxy, such as we hold today.[84] But whether the writings of Kant, or Wright, or J. H. Lambert[85] (who held similar views of the Galaxy, see Section B, chap. 4) were known to William Herschel in the early stages of his career is doubtful.[86] Herschel it was who not only realised (as his predecessors had done) that the Milky Way is the optical effect of our immersion in a layer of stars, but who proposed two assumptions and a technique, by means of which he would actually map the outline of the Galaxy.[87]

Herschel's first assumption was that his telescope could reach the borders of the Galaxy in every direction. Of this he had no proof, but it was a *necessary* assumption if he was to have a hope of solving the problem he had set himself, since if he could not even see to the edge of our Galaxy, he obviously could not map the Galaxy.

His second assumption was that, within the volume of space occupied by the Galaxy, the stars are situated at approximately regular intervals. This means that, the more stars we can see in a given direction, the farther it is to the border of the Galaxy in that direction. Again, Herschel had no proof whatever. Indeed, in particular instances it was in plain contradiction with his usual assumption

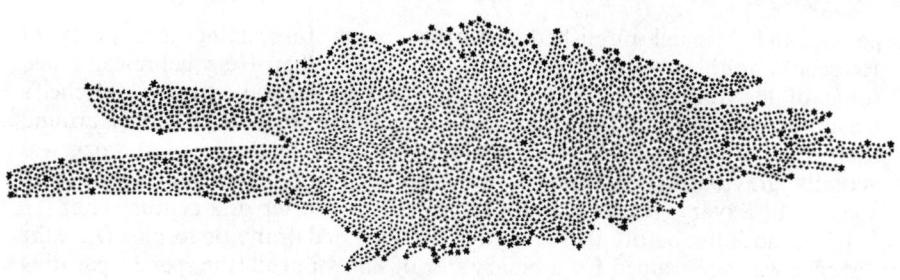

Fig. 1. William Herschel's cross-section of the Milky Way.

that, the fainter a star, the more distant it is.[88] Perhaps it is the mark of a truly great cosmologist, to work simultaneously with three assumptions, for none of which has he any evidence whatever, and of which two lead to contradictory consequences!

Herschel's technique was a pioneer exercise in stellar statistics. He counted the number of stars in neighbouring fields of view, took the average, and converted the result into a (relative) distance. The time he could spare from his regular observations permitted him to carry through this programme only for one great circle of the sky, and from this resulted (see Figure 1) his famous cross-section of the Milky Way (1785). But completion in 1789 of his monster telescope of 40ft focal length revealed many stars previously invisible, and so proved the falsity (in 1785, at least) of his first assumption. Meanwhile, his prolonged investigations of star clusters brought home to him how very far from the truth was his second assumption, of the uniform distribution of stars. His attempt to map the Galaxy had failed, but it had demonstrated the power of the technique of stellar statistics.

The nineteenth century saw relatively little progress in this subject, and indeed the cross-section drawn by Herschel continues to recur in textbooks, even though its author had disowned it. Wilhelm Struve in 1847 attempted an analysis of the star counts of Herschel and the catalogues of Bessel and Argelander, and proposed a model of the Galaxy according to which there is in the universe an infinite central plane throughout which the stars are scattered with a uniformly high degree of condensation, while on either side of this central plane the stars are scattered with a condensation that diminishes sharply with increasing distance from the central plane.[89] According to Struve, for stars at very great distances so much light is lost in its journey from the star to us that the star is invisible.[90]

This last assumption, and more generally the highly-artificial nature of the model, repelled such observers as John Herschel, who had pioneered the observation of the southern sky and who knew from personal experience just how complex and irregular are the meanderings of the Milky Way.[91] Herschel was convinced that in some directions he had seen through the Galaxy out into empty space, and that in such directions the Galaxy was clearly of limited extent. He inclined to the view that the Milky Way is an annulus or ring of small stars surrounding the solar system.[92] Other astronomers later in the century, such as R. A. Proctor,[93] Hugo von Seeliger[94] and J. C. Kapteyn,[95] modified this to allow more complicated shapes, notably elaborate spirals (by analogy with spiral nebulae). The difficulty was that William Herschel's criterion of the uniform distribution of the stars of the Galaxy – a dramatically simple and effective weapon until destroyed by opposing evidence – had to be replaced by a more fine-grained approach. It was necessary to give separate consideration to the distribution of the stars of different magnitudes; and this at a time when it was becoming increasingly clear that apparent brightness, still the only readily-available clue to distance, was a most unreliable guide. So the nineteenth century ended knowing little more than William Herschel's star-counts had shown in the 1780s, that the plane of the Milky Way contains more stars than are to be found in other parts of the sky.

REFERENCES

1. James Gregory, *Geometriae pars universalis* (Padua, 1668), 148. Gregory uses 3' for the solar parallax, and assumes *all* the light is reflected from Jupiter.
2. Christiaan Huygens, *Cosmotheoros* (Hagae-Comitum, 1698), 136-7.
3. In particular, by J. P. L. de Chéseaux (*Traité de la comète qui a paru en 1743 et 1744* (Lausanne and Geneva, 1744), 223-9); J. H. Lambert (*Photometria* (Augsberg, 1760), 504-11); John Michell ("An enquiry into the probable parallax and magnitude of the fixed stars", *Philosophical transactions*, lvii (1767), 234-64); and H. W. M. Olbers ("Mars und Aldebaran", *Monatliche Correspondenz*, viii (1803), 293-311). All derive values of several hundred thousand astronomical units.
4. Discussed in M. A. Hoskin, *William Herschel and the construction of the heavens* (London, 1963). This work is hereafter cited as *Construction*.
5. W. Herschel, "Account of the changes that have happened during the last twenty-five years, in the relative situation of double stars", *Philosophical transactions*, xciii (1803), 339-82.
6. Hoskin, *Construction*, 40.
7. Most notably in Cambridge University Library MS Add. 3965, f280^r; see Section B, chap. 2 below. Robert Hooke, *An attempt to prove the motion of the Earth by observations* (London, 1674), 14, suggests that "the Stars are further and further removed from the Sun, according as they appear less and less to us", and if we examine his diagram we see that a fourth magnitude star is taken to be four times further than one of the first magnitude. Similarly, in the Third Day of the *Dialogo*, Galileo has Salviati discuss the consequences of supposing a fixed star of the sixth magnitude to be really equal to the Sun, and tells us that "the apparent diameter of a fixed star of the first magnitude is no more than five seconds, that is to say 300 thirds, and the diameter of a fixed star of the sixth magnitude, 50 thirds" (Galileo, *Le Opere* (ed. by A. Favaro (Florence, 1890-1909)), vii, 386).
8. The assumption was challenged by Nevil Maskelyne after Herschel's paper "On the parallax of the fixed stars" was first read to the Royal Society, and Herschel beat a tactical retreat by claiming to be defining new magnitudes which would meet his requirements. See Hoskin, *Construction*, 32-35.
9. Cambridge University Library MS Add. 3965, f74^r; see Section B, chap. 2 below.
10. W. Herschel, "Astronomical observations, and experiments tending to investigate the local arrangement of the celestial bodies in space", *Philosophical transactions*, cvii (1817), 302-31. *Cf.* Hoskin, *Construction*, 171-4.
11. Herschel, *ibid.*, Section V; *cf.* Hoskin, *Construction*, 168. The method of measuring starlight by limiting apertures had been proposed in 1771 by J. S. Bailly in "Mémoire sur les inégalités de la lumière des satellites de Jupiter", *Mémoires de l'Académie royale des Sciences* for 1771, 580-667. The effect of different apertures is discussed by Michell, *op. cit.* (ref. 3), 259.
12. J. F. W. Herschel, *Results of astronomical observations made at the Cape of Good Hope* (London, 1847), 353-7.
13. "Observations with the meridian photometer", *Annals* of the Harvard Observatory, xiv, part 1 (1884); *Potsdamer Photometrische Durchmusterung* (Potsdam, 1894-1907); *Uranometria nova Oxoniensis* (Oxford, 1885).
14. Michell, *op. cit.* (ref. 3), has a good assessment of the importance of trigonometrical measurements of annual parallax.
15. Hooke, *op. cit.* (ref. 7).

16. Published in John Wallis, *Opera mathematica*, iii (London, 1699), 705-8. Flamsteed's blunder was quickly pointed out by J. D. Cassini, "Reflexions sur une lettre de M. Flamsteed à M. Wallis touchant la parallaxe annuelle de l'étoile polaire", *Mémoires de l'Académie royale des Sciences pour 1699* (1702), 177-83. It is explained in M. E. W. Williams, "Flamsteed's alleged measurement of annual parallax for the Pole Star", *Journal for the history of astronomy*, x (1979), 102-16.
17. James Bradley, "An account of a new discovered motion of the fixed stars", *Philosophical transactions*, xxxv (1727-28), 637-61.
18. James Bradley, "An apparent motion observed in some of the fixed stars", *Philosophical transactions*, xlv (1748), 1-43.
19. Bradley, "An account", 660.
20. In *A treatise of the system of the world* (London, 1728).
21. Full and critical analyses of these and other measurements are contained in two major studies: "Recherches sur la parallaxe des étoiles fixes" by C. A. F. Peters, *Mémoires de l'Académie des Sciences de Saint-Pétersbourg*, sér. vi, Sciences mathématiques et physiques, v (1853), 1-65, and the historical "Responsio" of G. R. Fockens on annual parallax, *Annales Academiae Lugduno-Bataviae* for 8 February 1834 to 9 February 1835 (Leiden, 1835), 1-263. For Piazzi's results, see "Ricerche di Giuseppe Piazzi su la parallasse annua di alcune delle principali fisse", *Memorie della Societa Italiana delle Scienze*, xii (Modena, 1805), 40-61.
22. G. Calandrelli, *Opuscoli astronomichi di Giuseppe Calandrelli e Andrea Conti* (Rome, 1806).
23. Early discussions are by J.-B. Delambre in "Histoire de l'astronomie pour 1804 et 1805", in *Connaissance des temps pour 1808* (Paris, 1806), 432-5, and by F. von Zach in "Resultate der neuesten Untersuchungen über jährliche Parallaxe der Fixsterne", *Monatliche Correspondenz*, viii (1808), 401-16; ix (1809), 38-60, 183-9, 234-53.
24. The relevant papers by Brinkley are conveniently listed in *Dictionary of scientific biography*, ii (New York, 1970), 469.
25. John Pond, "On the parallax of the fixed stars", *Philosophical transactions*, cvii (1817), 158-75; "On the parallax of the fixed stars", *ibid.*, cvii (1817), 353-62; "On the parallax of α Aquilae", *ibid.*, cviii (1818), 477-80; "On the parallax of the fixed stars in right ascension", *ibid.*, cviii (1818), 481-5; "On the changes ... in the declination of some of the principal fixed stars", *ibid.*, cxiii (1823), 34-52; "On the parallax of α Lyrae", *ibid.*, 53-72.
26. Edmund Halley, "Considerations on the change of the latitudes of some of the principal fixt stars", *Philosophical transactions*, xxx (1717-19), 736-8.
27. As the number of known proper motions grew in the later years of the eighteenth century, so astronomers became increasingly aware of large proper motion as a clue to nearness. For example, John Michell: "The apparent change of situation, that has been observed amongst a few of the stars, is a strong circumstance in favour of those stars being some of the nearest to us" (*op. cit.* (ref. 3), 252); Thomas Hornsby: "As none of the other principal stars have been found to have a motion so considerable as this [of Arcturus], ... we may, I think, fairly conclude that Arcturus is the nearest star to our system, visible in this hemisphere. If therefore the annual parallax of the fixed stars can ever be discovered, ... it is most likely to be discovered from the observations of Arcturus" ("An enquiry into the quantity and direction of the proper motion of Arcturus", *Philosophical transactions*, lxiii (1773-74), 93-125, p. 105); Edward Pigott: "...how far the discovery of proper motions in small stars may induce Astronomers to choose them for finding a parallax I dare not venture to decide" (writing to William Herschel in 1783, see chap. 4 below, ref. 11). *Cf.* John Brinkley, *Elements of plane astronomy* (Dublin, 1808), 127-8.
28. F. W. Bessel, "Über den Doppel-Stern Nro. 61 Cygni", *Monatliche Correspondenz*, vi (1812), 148-63. Piazzi pointed out plaintively that he had already remarked on the double star's large motion: "Harum stellarum motus proprii jam ab anno 1804 a nobis explorati, & in libro VI (Del Reale Osservatorio di Palermo) edito anno 1806 consignati" (G. Piazzi, *Praecipuarum stellarum inerrantium positiones mediae* (Palermo, 1814)); but it is one of three stars mentioned only in a footnote (p. 10).
29. See pp. 281-2 of "Notices scientifiques" by F. Arago in *Annuaire pour l'an 1834* (Paris, 1833).
30. Baron von Lindenau, "Beobachtung und Elemente der Bahn des Kometen von 1815; Ueber 61 im Schwan...", *Berliner astronomisches Jahrbuch für das Jahr 1818* (Berlin, 1815), 244-51.
31. Michael Hoskin, "Stellar distances: Galileo's method and its subsequent history", *Indian journal of history of science*, i (1966), 22-29.
32. Bradley, "An apparent motion" (ref. 18), 41. Others include James Gregory in a letter to Henry Oldenburg, 8 June 1675 (see H. W. Turnbull, *James Gregory tercentenary memorial volume* (London, 1939), 306-7); Newton (see ref. 20); John Wallis, "A proposal concerning the parallax of the fixed stars, in reference to the Earth's annual orb", *Philosophical transactions*, xvii (1693), 844-9, p. 848; and Huygens, *Cosmotheoros*, 134-5.

33. Hoskin, *Construction*, 31.
34. *Königsberger Beobachtungen*, ii (1816), p. VIII and iii (1817), p. IX. For Struve's explanation of Bessel's results, see Peters, "Recherches" (ref. 21), 49-51.
35. *Observationes Dorpatenses*, iii (Dorpat, 1822), pp. LI-LXXXX.
36. F. G. W. Struve, *Mensurae micrometricae* (St Petersburg, 1837), pp. CLXII-CLXXIII.
37. *Ibid.*, pp. CLXX-CLXXII.
38. F. G. W. Struve, *Additamentum in mensuras micrometricas* (St Petersburg, 1840), 20-30; "Über die Parallaxe des Sterns α Lyrae", *Astronomische Nachrichten*, no. 396 (1840), cols 177-80.
39. "A letter from Professor Bessel to Sir J. Herschel, Bart., dated Königsberg, Oct. 23, 1838", *Monthly notices of the Royal Astronomical Society*, iv (1839), 152-61.
40. *Ibid.*, 152.
41. R. Main, "On the present state of our knowledge of the parallax of the fixed stars", *Memoirs of the Royal Astronomical Society*, xii (1842), 1-60; J. W. F. Herschel, "Address to the Royal Astronomical Society on 12th February, 1841", *ibid.*, 442-54, reprinted in *Essays* (London, 1857), 532-51.
42. Otto Struve, "The first stellar parallax determination", in Herbert M. Evans (ed.), *Men and moments in the history of science* (Seattle, 1959), 177-206.
43. T. Henderson, "On the parallax of α Centauri", *Memoirs of the Royal Astronomical Society*, xi (1840), 61-68. The value he obtained was $1''\cdot 16$, somewhat too large.
44. On the progress of the subject during the next half-century, see Agnes M. Clerke, *The system of the stars* (London, 1890), chap. 20: "The distances of the stars".
45. John Flamsteed, *Historia coelestis Britannica* (London, 1725).
46. *Astronomical observations made at the Royal Observatory at Greenwich from the year 1750 to the year 1762 by the Rev. James Bradley* (2 vols, Oxford, 1798-1805).
47. F. W. Bessel, *Fundamenta astronomiae pro anno 1755* (Königsberg, 1818).
48. A. von Auwers, *Neue Reduktion der Bradley'schen Beobachtungen 1750 bis 1762* (3 vols, St Petersburg, 1882-1903).
49. David Gill, "On photographs of the great comet (b) 1882", *Monthly notices of the Royal Astronomical Society*, xliii (1882-83), 53-54.
50. David Gill and J. C. Kapteyn, "The Cape photographic Durchmusterung", *Annals of the Cape Observatory*, iii (1896) – v (1900).
51. See G. Van Biesbroeck, "The *Carte du ciel* catalogue", in K. Aa. Strand (ed.), *Basic astronomical data* (Chicago, 1963), 474-7.
52. His first "Catalogue of the comparative brightness of the stars" appeared in "On the method of observing the changes that happen to the fixed stars", *Philosophical transactions*, lxxxvi (1796), 166-226. E. C. Pickering, a century later, declared that "Herschel furnished observations of nearly three thousand stars, from which their magnitudes a hundred years ago can now be determined with an accuracy approaching that of the best modern catalogues" (*Harvard annals*, xxiii (1890), 231). In 1767, Michell (*op. cit.* (ref. 3)) asked that the stars "may be ranked with precision both according to their respective brightness, and the exact degree of it" (p. 241). "It would", he adds, "be a standing register, by which future Astronomers might inform themselves of many variations, of which we are now ignorant for want of an ancient register of that kind".
53. Ismael Bullialdus, *Ad astronomos monita duo* (Paris, 1667), 5-14.
54. *Ibid.*, 14.
55. See the discussion of this and other new and variable stars in the fifth of William Whiston's *Praelectiones astronomicae* (Cambridge, 1707), translated as *Astronomical lectures* (London, 1715, 1728).
56. See for example Newton's remarks, Cambridge University Library MS Add. 4005, ff21-22, published in A. R. Hall and M. B. Hall, *Unpublished scientific papers of Sir Isaac Newton* (Cambridge, 1962), 376.
57. See for example James Ferguson, *Astronomy explained upon Sir Isaac Newton's Principles* (2nd edn, London, 1757), §367.
58. As suggested by P. L. M. de Maupertuis, *Discours sur les différentes figures des astres* (Paris, 1732).
59. John Goodricke, "A series of observations on ... Algol", *Philosophical transactions*, lxxiii (1783), 474-82.
60. For further information see Section A, chap. 4 below and the detailed discussions in *Astronomie* by J.-J. de Lalande (3rd ed., Paris, 1792), i, 259-68, and in *Popular astronomy* by D. F. J. Arago (Paris, 1855), i, 239-61.

61. Halley, "Considerations on the change of the latitudes of some of the principal fixt stars" (ref. 26).
62. *The posthumous works of Robert Hooke* (London, 1705), 506, written before 1669; William Whiston, *Sir Isaac Newton's mathematick philosophy more easily demonstrated* (London, 1716), 349.
63. Bradley, "An apparent motion observed in some of the fixed stars" (ref. 18), 40. See also Thomas Wright of Durham, *An original theory of the universe* (London, 1750), 52, 55.
64. *Opera inedita Tobiae Mayeri, I* (Göttingen, 1775), 77-81; trans. by Eric G. Forbes, *Tobias Mayer's "Opera inedita"* (London, 1971), 108-12.
65. William Herschel, "On the proper motion of the Sun and solar system", *Philosophical transactions,* lxxiii (1783), 247-83.
66. P. Prévost, "Mémoire sur le mouvement progressif du centre de gravité de tout le système solaire", *Nouveaux mémoires de l'Académie royale des Sciences et Belles-lettres, Année MDCCLXXXI* (Berlin, 1873), 418-21; G. S. Klügel, "Trigonometrische Formeln zu der Untersuchung über die Fortrückung der Sonne und der Sterne", *Berliner astronomisches Jahrbuch für das Jahr 1789* (Berlin, 1786), 214-24.
67. William Herschel, "On the direction and velocity of the motion of the Sun, and solar system", *Philosophical transactions,* xcv (1805), 233-56, and "On the quantity and velocity of the solar motion", *ibid.,* xcvi (1806), 205-37.
68. Hoskin, *Construction,* 55-59.
69. Bessel, *Fundamenta,* 308-13.
70. F. Argelander, "Über die eigene Bewegung des Sonnensystems", *Mémoires présentés à l'Académie impériale des Sciences,* iii (St Petersburg, 1837), 561-605.
71. Michell, "An inquiry into the proper parallax, and magnitude of the fixed stars" (ref. 3). Michell's work is discussed by Clyde L. Hardin, "The scientific work of the Reverend John Michell", *Annals of science,* xxii (1966), 22-47, and Russell McCormmach, "John Michell and Henry Cavendish: Weighing the stars", *British journal for the history of science,* iv (1968), 126-55.
72. *Ibid.,* 246.
73. John Michell, "On the means of discovering the distance, magnitude etc, of the fixed stars", *Philosophical transactions,* lxxiv (1784), 35-57.
74. William Herschel, *op. cit.* (ref. 5).
75. F. Savary, "Sur la détermination des orbites que décrivent autour de leur centre de gravité deux étoiles très-rapprochées l'une de l'autre", *Connaissance des temps ... pour l'année 1830* (Paris, 1827), 56-69, 163-171; J. F. Encke, "Über die Berechnung der Bahnen der Doppelsterne", *Berliner astronomisches Jahrbuch für das Jahr 1832* (Berlin, 1830), 253-304; J. F. W. Herschel, "On the investigation of the orbits of revolving double stars", *Memoirs of the Royal Astronomical Society,* v (1833), 171-222.
76. In a letter published in *Monthly notices of the Royal Astronomical Society,* vi (1845), 136-41.
77. See the account in T. H. Safford, "On the observed motions of the companion of Sirius", *Proceedings of the American Academy of Arts and Sciences,* vi (1866), 143-6.
78. While testing the $18\frac{1}{2}''$ objective later installed in the Dearborn Observatory in Chicago.
79. By J. M. Schaeberle at Lick Observatory, 13 November 1896.
80. Clerke, *The system of the stars,* 172-3.
81. S. L. Jaki, "The Milky Way before Galileo", *Journal for the history of astronomy,* ii (1971), 161-7; and *The Milky Way* (New York, 1972), which has an extended discussion of the material summarized in what follows.
82. Wright, *An original theory.* A reprint of this work, with transcription of related manuscripts, has been edited by M. A. Hoskin (London, 1971).
83. *An original theory,* 63.
84. I. Kant, *Allgemeine Naturgeschichte und Theorie des Himmels* (Königsberg and Leipzig, 1755), translated by W. Hastie as *Kant's cosmogony* (Glasgow, 1900; reprinted Ann Arbor, 1969, and New York, 1970).
85. J. H. Lambert, *Cosmologische Briefe* (Augsburg, 1761), translated by S. L. Jaki under the title *Cosmological letters on the arrangement of the world-edifice* (Edinburgh and New York, 1976).
86. On the possibility of Herschel's early familiarity with the work of Wright, see Section B, chap. 3 below; on his familiarity with Lambert, see Section B, chap. 4.
87. William Herschel, "On the construction of the heavens", *Philosophical transactions,* lxxv (1785), 213-66; Hoskin, *Construction,* chap. 3.

88. Hoskin, *Construction*, 65.
89. F. G. W. Struve, *Études d'astronomie stellaire* (St Petersburg, 1847).
90. *Ibid.*, 86.
91. J. F. W. Herschel, *Outlines of astronomy* (London, 1849), para. 797-8.
92. *Ibid.*, para. 788.
93. R. A. Proctor, "A new theory of the Milky Way", *Monthly notices of the Royal Astronomical Society*, xxx (1869-70), 50-56.
94. In a series of papers presented to the Munich Academy from 1884.
95. An account of Kapteyn's researches is conveniently available in *Dictionary of scientific biography*, vii (New York, 1973), 235-40. A useful survey by Kapteyn himself is "Recent researches in the structure of the universe", Friday Evening Discourse at the Royal Institution, London, 22 May 1908 (*Proceedings of the Royal Institution of Great Britain*, xix (1908-10), 300-15, reprinted in *The Royal Institution library of science: Astronomy*, ii (London, 1970), 78-96).

2. *Novae and Variables from Tycho to Bullialdus*

The modern astronomer regards every star as variable in brightness to a greater or less extent: at one extreme, mankind is almost unaware of the unobtrusive but significant changes in our own Sun such as those reflected in the sunspot cycle; at the other, the nuclear explosion of a supernova may increase the luminosity of the star by a factor of hundreds of millions in a matter of hours. In complete contrast, in the mid-sixteenth century the stars were regarded as "fixed", unchanging both in brightness and in position, as natural philosophy demonstrated and records since the time of Hipparchus confirmed. The first serious doubts concerning the possibility of change in the heavens were aroused, as is well known, in November 1572 by the appearance in the constellation of Cassiopeia of what appeared to be a very bright new star. It took astronomers another century to glimpse something of the frequency and the bewildering variety of changes among the stars. The story of this century, the details of which are often misunderstood,[1] is the subject of this chapter.

Despite the universality of change among the stars as established by modern astronomy, we need not be surprised at the failure of the mid-sixteenth-century observers to recognise any such alterations. The theoretical contrast between the changing Earth and the unchanging sky was apparently supported by astronomical evidence going back to Antiquity. For of the naked-eye stars listed as 'variable' by astronomers today, few fluctuate in brightness by more than a couple of magnitudes (bright stars being classified since Ptolemy as first magnitude and the faintest visible to the unaided eye as sixth). It was only *after* the appearance of the 1572 nova that Tycho Brahe opened the modern era of precision astronomy by his long campaign for observational accuracy; and even so, the fact that Tycho had left no record of a (relatively bright) star of the third magnitude would not necessarily be accepted as clear proof that in Tycho's day the star was not to be seen in the sky.[2] It therefore should not surprise us that before Tycho no variable or new star was known in the West; or that the trumpet call of a supernova, rising with dramatic suddenness to rival Venus and be visible even in daytime, should have been necessary to draw attention to the changes that occur in the heavens.

Such supernovae, it is now believed, occur somewhere in our Galaxy once every few tens or hundreds of years. Before that of 1572, the last of which we have knowledge had appeared in the eleventh century, when European astronomy was still in a very primitive condition.[3] The intervening centuries had seen substantial improvements in the accuracy of observation;[4] and it so happened that "Tycho's nova" appeared at a time when astronomical skills were at last adequate to the simple but by no means trivial task of proving by measurement that the new appearance was indeed in the sky and not in the atmosphere.[5] Not everyone was agreed on this, and some observers found otherwise, but the two works which Tycho wrote on the nova,[6] supported by his fame as an observer, convinced the best astronomers of the next generation. By a very remarkable coincidence, although the nova of 1572 gradually declined in brightness and vanished from sight in the spring of 1574, Kepler's generation could make their own measurements when a very similar supernova, only slightly less bright, appeared in 1604, in the constellation of Ophiucus; "Kepler's nova" took a year to decline and vanish from sight.[7]

Both supernovae were distant stars and so were observed close to the galactic equator, that is, near the Milky Way; and the Milky Way was quickly recognised as "the fund of new stars",[8] a fact which influenced physical theories of the origin of new stars.[9] There were, in addition, two remarkable coincidences, the first resulting from the discussion of the astronomer Cyprianus Leovitius in his pamphlet on the 1572 star published the following year. The apparition in Cassiopeia naturally led to a search of old records in the hope of finding previous examples of sightings of new stars, beardless comets and the like; for one thing, the astrological significance of new stars might be better understood as a result of such a comparative study. Leovitius found records which he believed showed that stars had appeared in the same region of the heavens in 954 and 1264, and Tycho relates the substance of Leovitius's remarks, which thereby passed into the literature. After periodic variables were recognised, at the end of the century with which we are concerned, it was noticed that the interval between 945 and 1264 is almost exactly that between 1264 and 1572, which led astronomers to ask whether this might be the same star appearing every three (or one-and-a-half) centuries – a possibility of much interest to the late nineteenth century when the star might be expected to reappear.[10]

The other coincidence concerns Kepler's nova, which appeared near the place of conjunction of the planets Mars, Jupiter and Saturn at a time of exceptional astrological significance: the beginning, late in 1603, of the eight-hundred-year period of the fiery trigon, most distinguished of the trigons defined by conjunctions of Jupiter and Saturn. This extraordinary chance offered endless scope for astrological speculation.[11]

It is revealing of the difficulties surrounding this subject at the turn of the century that, in the interval between these two indisputable appearances and disappearances, there were three other European reports of new stars, and each was of questionable authenticity. One of these 'stars' was glimpsed briefly by Kepler in 1602 while he was observing the Moon, but he himself was doubtful whether he had indeed seen a new star.[12]

A second had supposedly appeared in 1596, in the constellation of the Whale, far from the Milky Way. It was reported only by David Fabricius in Frisia, as a star of the third magnitude. He first noticed it in August, and it vanished again in October. A sighting by a single observer might not normally have carried conviction, but Fabricius was, after Tycho, the finest observer of his day and much respected by both Tycho and Kepler. He sent his observations to Tycho,[13] and Kepler reports the sighting briefly in his *Ad Vitellionem paralipomena*[14] and, with coordinates and at length, in his treatise on the 1604 star, *De stella nova*.[15] But that, it seemed, was the end of the matter; not all astronomers believed that the star had ever existed.[16]

That the third star, in the Swan, existed was beyond doubt, for it remained in the sky for many years. It was first noted in 1600 by Gulielmus Jansonius

(Willem Janszoon Blaeu, the great cartographer) and inserted by him on a celestial globe the following year.[17] Blaeu had been a pupil of Tycho and was much respected by Kepler; but Kepler's letter asking how he had noticed the new star went unanswered.[18] Kepler discusses with care whether this third magnitude star, clearly visible to all but situated in an uninteresting region of the sky, could have been overlooked by previous astronomers, and concludes that it is genuinely new.[19] Riccioli, writing in the mid-century with the knowledge that the star had been visible for at least two decades and that it disappeared sometime in the 1620s (though exactly when was contested[20]), accepts it with the stars of 1572 and 1604 as the only authentic novae since Hipparchus.[21]

There were two reports of novae in 1612 and another in 1618. One of the 1612 'novae' was in fact the Andromeda Nebula, discovered with a telescope by Simon Marius on 15 December; the other, ascribed to Joost Bürgi, was said to have appeared in what was then the constellation of Antinous.[22] The 1618 report came from the observer of the comet of that year while on board ship for the East Indies, but he himself considered the sighting doubtful.[23] Then, in 1638, at the time of an eclipse, the Frisian astronomer Johannes Phocylides Holwarda observed in the Whale a star of better than third magnitude which was listed neither by Hipparchus and Ptolemy nor by Tycho and therefore, he insists, must be new.[24] Holwarda reminds his readers of the new stars discussed by Kepler, including that in the Whale claimed by Fabricius.[25] It is significant that it does not occur to Holwarda that his star might be a reappearance of Fabricius's nova, for Fabricius's coordinates, when corrected for precession, give a position within a few minutes of Holwarda's.[26]

The 1638 nova, it seemed, had nothing to mark it off from the earlier and more dramatic new stars, and Holwarda prepared his long but routine account of its discovery and eventual disappearance. We can imagine his amazement when, after the text of his book had been printed off, the star reappeared. The exciting news is added in an *Appendix ad Lectorem necessaria;*[27] but even then Holwarda does not identify his star with that of Fabricius, nor with the o Ceti of Bayer's 1603 celestial charts. Nevertheless, although there had been many speculations that 'novae' might be existing stars which had increased in brightness, this was the first known example of a star reaching more than one maximum of brightness.

In the 1640s, the star's fluctuations were observed from time to time, by Fullenius and Jungius, and on 5 January 1648 by Hevelius himself. But Hevelius was diverted to other matters, and only came back to the star – which, following Jungius, he named "Mira", the wonderful star – in August 1659. He then followed the star closely for three years before publishing details of its fluctuations in his *Historiola mirae stellae* (1662).[28] It is in this work that he remarks on its proximity to Fabricius's nova and to o Ceti.

Because his observations were effectively confined to only three years, Hevelius failed to detect a rhythm which underlay the wild fluctuations in Mira's brightness at successive maxima. This important discovery was therefore left for Ismael Bullialdus who, after a further three years, noticed that the date of Mira's maximum brightness was occurring a month earlier each year.[29] By going back nearly thirty years to Holwarda's observations in 1638, Bullialdus was able to specify a periodicity of some 333 days, a very accurate value. For the very first time, a changing star was lawlike in its behaviour, and Bullialdus could actually invite astronomers to observe future maxima on dates which he could announce to them in advance; prediction was possible, and to that extent understanding had been achieved. He also offered a physical explanation of variable stars that was to endure for centuries. As the review in the *Philosophical transactions* put it:

> ... the bigger part of that round Body is obscure and inconspicuous to us, and its lesser part lucid, the whole Body turning about its own Center, and one Axe; whereby for one determinate space of time it exhibits its lucid part to the Earth, for another, subducts it: it not being likely, that fires should be kindled in the Body of that Star, and that the matter thereof should at certain times take fire and shine, at other times be extinguisht upon the consumption of that matter.[30]

The study of variable stars was evidently fruitful as well as technically undemanding; news of discoveries could be quickly disseminated through the infant Royal Society or Académie des Sciences,[31] and variable stars soon acquired the fascination for amateur astronomers which they hold to the present day. The next star to behave surprisingly was the nova of 1600 "in pectore Cygni" which Hevelius saw in 1658 as a star of the third magnitude but which vanished in 1662; in 1665 he saw it again as a star of the sixth or seventh magnitude.[32] The same year, Bullialdus rediscovered the Andromeda Nebula, which had not been seen since the time of Marius and so must surely be of variable brightness.[33] Then, in 1667, Geminiano Montanari, professor of mathematics at Bologna, noticed that the star Algol (β Persei) was of only fourth magnitude; in 1669 it was of second magnitude, and in 1670 again of the fourth.[34] On 20 June 1670, a Carthusian of Dyon named Voituret Anthelme, who had been keeping a special watch for new stars in the constellations of the Swan and Cassiopeia, found a new star "circa et infra caput Cygni" of the third magnitude. In Paris Cassini and others kept a close watch on the star during the summer and saw it slowly decline. During the winter it was invisible, but in March Dom Anthelme noticed it again, and it reached a maximum on 4 April and again at the beginning of May:

> As far as can be judged from the few Observations made of this Star, 'tis likely she is returning about *Ten* months unto the same appearance; whereas that in the *Whale's Neck* [Mira] maketh its revolution in *Eleven* months. As for the Star in the *Swans Breast* [the star of 1600], we have as yet no certain knowledge of the period of her revolution; yet one may assure, that she taketh no less than *Fourteen years* to finish it.[35]

By the centenary of the appearance of Tycho's nova, it was evident how widespread are the changes in the magnitudes of the stars. Indeed, the wheel had now turned full circle, and innumerable fictitious changes were being announced on all sides. Montanari had written to the Royal Society in April 1670 with an account of some of his discoveries, adding: "I have observed many more changes among the Fixed Stars, even to the Number of a Hundred".[36] And the *Journal des sçavans* reported that "Monsieur *Cassini* hath discover'd many other little [stars], which may very well be presumed to be New".[37] The stars which had aroused special interest were indeed representative of the enormous variety among the new and variable stars: those of 1572 and 1604 were true supernovae; that of 1600 was a 'shell star' giving off matter into space at intervals;[38] Dom Anthelme's star was a galactic nova;[39] Mira is the prototype of long-period intrinsic variables;[40] and Algol is a binary star whose members eclipse each other as they orbit in just under three days. The curtain had risen on the bewildering variety of phenomena that still intrigues and challenges the astronomer today.

REFERENCES

1. For example, the observation of Mira Ceti by Fabricius in 1596 is always cited, but historically this was of negligible consequence; by contrast, the new star of 1600 is seldom mentioned, and the apparent variability of the Andromeda Nebula is usually ignored. Again, Hevelius is often represented as keeping watch on Mira throughout the 1650s, in which case his failure to note its periodicity would be hard to explain.
2. As we shall see, it was argued whether or not the third-magnitude star of 1600 was genuinely new, and the same happened later with Mira Ceti. Even the supernova of 1572 was not as obvious as we tend to think. Tycho himself noticed it on 11 November, but it had certainly appeared several days before, probably between the 2nd and the 6th (see J. L. E. Dreyer's full discussion in his *Tycho Brahe* (Edinburgh, 1890), chap. 3, espec. 60-63) and possibly late in October. The enthusiastic astronomer Langrave Wilhelm IV did not hear of it until December; similarly, when Tycho visited Professor Johannes Pratensis in Copenhagen early in 1573, Pratensis knew nothing of the nova (Dreyer, *op. cit.*, 57, 42).
3. For discussions of novae in the pre-telescopic era, as revealed by historical records, see F. Richard Stephenson and David H. Clark, "Historical supernovas", *Scientific American*, ccxxxiv, no. 6 (June 1976), 100-7, and F. R. Stephenson, "A revised catalogue of pre-telescopic galactic novae and supernovae", *Quarterly journal of the Royal Astronomical Society*, xvii (1976), 121-38.
4. *Cf.* Olaf Pedersen and Morgens Pihl, *Early physics and astronomy* (London, 1974), where the role of the astrolabe in promoting accuracy of observation is emphasised.
5. Dreyer, *Tycho Brahe*, 38-41, 57-60.
6. Tycho Brahe, *De nova stella* (Copenhagen, 1573, reprinted in *Opera*, ed. by J. L. E. Dreyer (15 vols, Copenhagen, 1913-29), i), and his vast *Astronomiae instauratae progymnasmata* (Prague 1602, reprinted in *Opera*, ii and iii).
7. Johannes Kepler, *De stella nova in pede Serpentarii* (Prague, 1606).
8. The English phrase is used in the translation of David Gregory's discussion of new and variable stars (*The elements of physical and geometrical astronomy*, 2nd edition (London, 1726), i, 312).
9. It is not the purpose of this paper to discuss these theories. They are treated at great length by Tycho and Kepler in the works cited above and of course by innumerable other writers of the period, notably Fortunius Licetus, *De novis astris et cometis* (Venice, 1623), and J. B. Riccioli, *Almagestum novum* (Bologna, 1651/1653), ii, Liber VIII, Sectio II: De novis stellis. Riccioli gives a list of alleged new stars (chap. 1); discusses at length the stars of 1572 (chaps. 2-11), 1600 (chaps. 12-13) and 1604 (chaps. 14-15), with synopsis (chap. 16); considers their material, nature and generation (chap. 17) and efficient and final causes (chap. 18); and concludes with a discussion of the Star of Bethlehem (chaps. 19-20). He is as always a mine of information.
10. Cyprianus Leovitius, *De nova stella* (Lavingae, 1573). Leovitius quotes "historiae" as his source for the alleged 945 nova and takes the 1264 nova "ex antiquo codice, manu scripto" (see Dreyer, *Tycho Brahe*, 65 and Tycho's *Opera*, ii, 455 for the relevant sections *verbatim*); his claims became common knowledge when reported by Tycho in *Progymnasmata*, 331, 531, 706 (= *Opera*, ii, 328; iii, 46, 219) and so passed into the literature for centuries. John Keill, in his *Introductio ad veram astronomiam* (Oxford, 1718), 60, reports Leovitius's announcement of the stars of 945 and 1264, and continues: "It is probable that these two *Stars* might have been the same with that which was seen by *Tycho*". The text is unaltered in the second Latin edition (London, 1721), but in the English translation of that year (*An introduction to the true astronomy* (London, 1721), 57) he continues: "...and that in about 150 Years the same *Star* may again make its Appearance". This is the earliest suggestion of its kind known to me. On the

prospect of a sighting in the late nineteenth century, cf. Dreyer, *Tycho Brahe*, 56-67. On the 945 star, see the letter of W. T. Lynn on "The supposed comet of A.D. 945", *Observatory*, xviii (1895), 335-6. I wish to thank Dr W. H. Donahue for discussions on this problem.

11. See Kepler, *De stella nova*, and Max Caspar, *Kepler*, trans. by C. D. Hellman (New York, 1959), 153-6.
12. Kepler, *Ad Vitellionem paralipomena* (Frankfurt, 1604), 237; *Gesammelte Werke* (Munich, 1937-), ii, 209.
13. "Observationes quas misit mihi Dominus David Fabricius", Tycho Brahe, *Opera*, xiii, 114-15.
14. Kepler, *Ad Vitellionem paralipomena*, 446; *Werke*, ii, 376.
15. Kepler, *De stella nova*, 112; *Werke*, i, 259: "Prima, quod David Fabricius, quem in observationibus supra quoque fide dignum celebravi, animadvertit anno 1596. 3/13. Augusti... matutino tempore novam stellam tertiae magnitudinis invenit in 25. 45' Arietis, cum latitudine Australi 15. 54'; quae post Octobrem eiusdem anni disparuit. Ille vero locus a via lactea abfuit longissime".
16. Riccioli, for example, considers it doubtful.
17. Kepler, *De stella nova*, 164; *Werke*, i, 307.
18. *Ibid.*
19. His discussion of the star occupies pp. 149-68 of *De stella nova* (*Werke*, i, 293-311).
20. Riccioli, *Almagestum novum*, ii, 166, where he cites various dates in the 1620s. Riccioli naturally does not consider the possibility that the star disappeared and then reappeared. He does however affirm it was not visible in the 1640s although Hevelius and later writers sometimes ignore this specific statement.
21. *Ibid.*, 132.
22. Simon Marius, *Mundus Jovialis* (Nuremberg, 1614), Preface, 15; Riccioli, *op. cit.*, 132.
23. Riccioli, *op. cit.*, 132.
24. J. P. Holwarda, *Dissertatio astronomica* (Franeker, 1640), 200.
25. *Ibid.*, 202-3.
26. Holwarda's position for the nova is longitude 26°4' Aries, latitude 16°10'A (*ibid.*, 196).
27. *Ibid.*, 285. The text is given verbatim by J. Hevelius, *Cometographia* (Danzig, 1665-68), 376-7. Riccioli, who cites from Holwarda's book, may have overlooked the Appendix, for he has no knowledge of the reappearance of the star.
28. Pp. 146-71 of *Mercurius in Sole visus* (Danzig, 1662). Hevelius gives full details of the previous sightings by Holwarda, Fullenius and Jungius (who wrote of it as "mira stella", 149) and makes it clear (contrary to what is often said) that he himself made no systematic observations of Mira in the decade following 1648: "A quo vero tempore ad annum 1659, ut ingenue fatear, me illam raro admodum, quantum memini, data opera quaesivisse" (149). Of Bayer's o Ceti, he writes: "Non quidem ex eo potissimum, quod in Uranometria Beyeri, in Asterismo Ceti, ejusq; collo, sed potius juxta curvaturam, seu gibbum ad o, certam quandam quartae magnitudinis, tam in Catalogo Ptolemaico, quam Tychonico nusquam extantem, inveniamus. Nam haec non prorsus eo loco, ubi nostra mira sita est, apparet" (166-7). He also remarks on the observation of Fabricius (168).
29. Ismael Bullialdus, *Ad astronomos monita duo* (Paris, 1667): Monitum Primum, De Stella Mirabili quae in Collo Ceti conspicitur (5-14).
30. *Philosophical transactions*, i (1665-67), 382; cf. *Journal des sçavans* for 16 January 1667, 11-12. The analogy with sunspots made this explanation convincing, and it could be adapted to explain almost any changes. Riccioli (*op. cit.*, 177), writing in ignorance of the reappearance of any new stars, had already suggested that some stars are partly dark and partly bright, and by fiat of God or otherwise are caused to make a partial rotation and so are seen by us as novae.
31. The early volumes of the *Philosophical transactions* contain numerous items on new and variable stars.
32. See for example the letter from Hevelius, *Philosophical transactions*, i (1665-67), 349.
33. Bullialdus, *Ad astronomos monita duo*: Monita Alterum, De Nebulosa stella, in Cingulo Andromedae anno 1665 ineunte deprehensa (14-19). Bullialdus quotes *verbatim* from the 1612 observation of Marius. The possibility that nebulae are variable was to be at the heart of theories of nebulae until well into the twentieth century (see Section C).
34. Geminiano Montanari, "Sopra le sparizione d'alcune stelle et altre novità celesti", *Prose de' signori accademici Gelati* (Bologna, 1671), 369-92. Montanari announced other discoveries of changes among the stars in a letter to the Royal Society dated 30 April 1670, which was read at a

35. meeting on 27 October 1670 (Thomas Birch, *A history of the Royal Society of London* (London, 1756-57), ii, 448). The letter proved memorable: see ref. 36 below.
35. The account in the *Journal des sçavans* for 22 June 1671, 32-36, was translated in *Philosophical transactions*, vi (1671), 2198-202. The quotation is from pp. 2200-1.
36. See ref. 34 above. The Latin text is given in the margin, *ibid.*, 2202. It made a deep impression: Nicholas Mercator quotes Montanari's Latin in his *Institutionum astronomicarum libri duo* (London, 1676) and the English translation used here is from John Keill, *An introduction to the true astronomy*, 58.
37. *Philosophical transactions*, vi (1671), 2201, translated from *Journal des sçavans* for 22 June 1671, 35. Numerous examples of changes observed by Cassini are then cited.
38. The star P Cygni, on which see C. Payne-Gaposchkin and S. Gaposchkin, *Variable stars* (Cambridge, Mass., 1938), 266-7, and Mart de Groot, "Mass loss from P Cygni", in *Mass loss from stars*, ed. by M. Hack (Dordrecht, 1969), 26-35.
39. The star CK Vulpeculae, on which see C. Payne-Gaposchkin, *The galactic novae* (Amsterdam, 1957), 212.
40. Mira ceases to be a naked-eye star for about five of the eleven months of its cycle, while at maximum it can be as bright as second magnitude. It therefore is exceptionally conspicuous as a variable star.

3. *Hooke, Bradley and the Aberration of Light*

In the aftermath of Copernicus's *De revolutionibus* (1543), those who denied the motion of the Earth had a powerful argument in the continuing failure of observers to detect 'annual parallax', the apparent annual cyclic movements of the stars to be expected if the Earth was indeed carrying observers in orbit around the Sun. The Copernican response was that so far observers had used instruments inadequate to the delicate task; but even when, by the middle of the seventeenth century, it was clear that the Copernicans were winning the argument on other grounds, to a convinced Copernican like Robert Hooke, Curator of Experiments to the newly-founded Royal Society, the position over annual parallax was very unsatisfactory. Hooke therefore carefully analysed the problem, discussed it at the Royal Society, and in 1669 constructed a novel telescope designed to make observations to an accuracy never before attempted, and for the sole purpose of measuring the annual parallax of one specially-situated star.

The use of a telescope in precise observations was itself something of a novelty, and Hooke's great contemporary, the Danzig observer Hevelius, was strongly opposed to this. But "'tis hardly possible", Hooke points out, "for any unarmed eye well to distinguish any Angle much smaller then that of a minute".[1] This granted, there was the problem that in observations spread over many months, the apparatus might bend under its own weight or, more subtly, be affected by the annual cycle of the seasons in such a way as to produce phantom annual movements in the stars:

> For if the Instruments be made of Wood, 'tis manifest that moyst weather will make the frame stretch, and dry weather will make it shrink a much greater quantity then to vary a minute: and if it be Metal, unless it be provided for in the fabrick of the Instrument accordingly, the heat of Summer, when the Summer observations are to be made, will make the Quadrant swell, and the cold of Winter will make it shrink much more then to vary a minute: Both of which inconveniences ought to be removed.[2]

Next, there was the problem of atmospheric refraction, whose effects were poorly understood and all too likely to vary with the weather conditions. Hooke's solution was to observe only when a star was vertically overhead; and fortunately there was a suitably bright star, γ Draconis, that passed almost directly over Hooke's London lodgings. To observe this one star, Hooke devised a telescope of novel construction: one fixed in the vertical direction, and with a micrometer to measure the distance of γ Draconis from the zenith at meridian passage.

The construction is depicted in Figure 1. The object glass is mounted in an inner tube, and this slides in an outer tube rigidly fixed to the roof. The object glass therefore points always in the same direction, but can be adjusted for focussing. It is protected from the weather by a lid which can be tilted open by

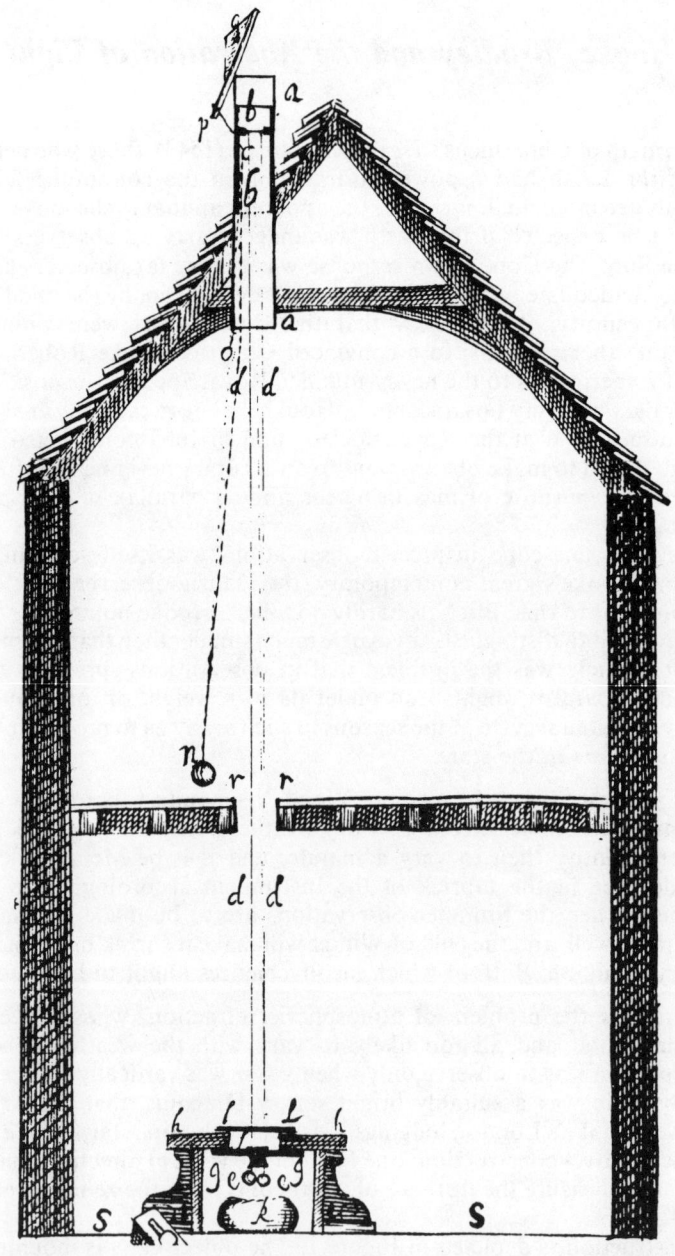

FIG. 1. Hooke's zenith telescope (from his *An attempt to prove the motion of the Earth* (London, 1674)).

pulling on a rope (shown tied slightly obliquely). In the floor of the upper room is a hole through which hang two vertical plumb lines some 36ft in length, suspended from either side of the centre of the object glass. Their ends pass through a hole in a table fixed to the ground floor, and they serve to locate a micrometer fixed on the top of the table. There is no conventional tube, and the eyepiece slides *horizontally* in a holder on the underside of the table; on the floor itself is a couch on which Hooke would lie. By moving the eyepiece sideways Hooke could bring fainter stars into view to give himself warning of the approach of γ Draconis, whose distance from the zenith at meridian passage he would then measure with the micrometer.

This "Archimedean Engine that was to move the Earth"[3] was ingenious in conception and cunning in execution. But it was a disappointment to its maker:

> ...I was forced to adjust the Instrument every observation I made, both before and after it was made, which hath often made me wish that I were near some great and solid Tower, or some great Rock or deep well ... for I often found that when I came to examine the Instrument, a day, or two, or three, or more, after a former observation, that there had been wrought a considerable change in the Perpendiculars, in so much as to vary above a minute from the place where I left them, which I ascribe chiefly to the warping of the Tube that rose above the roof of the House, finding sensibly that a warm day would bend it considerably towards the South, and that a moist Air would make it bend from the quarter of the wind: But yet I am apt to think there might be somewhat also of that variation ascribable to the whole Fabrick of the Roof, and possibly also to some variation of the Floors ....[4]

Worse was to follow, for when Hooke had made only four observations, in July, August and October 1669, "inconvenient weather and great indisposition in my health, hindred me from proceeding any further with the observation that time";[5] and later the object glass was accidentally broken.[6]

Nevertheless, from these four observations Hooke claimed to derive a (double) parallax of some 27″ or 30″, and "consequently a confirmation of the *Copernican* System against the *Ptolomaick* and *Tichonick*".[7] Needless to say, a parallax derived from so few observations seemed unconvincing to his contemporaries,[8] and Hooke himself urges other astronomers to learn from his mistakes; while later investigations have shown that Hooke's readings were indeed seriously flawed.[9] On the other hand, the next generation of astronomers had no way of knowing this, and half a century later, after all other attempts on annual parallax had failed, it seemed to the gifted amateur astronomer Samuel Molyneux that the time had come to revive Hooke's method.[10] For this he enlisted the help of James Bradley, professor of astronomy at Oxford, who as a young man had been initiated into observational techniques by his uncle James Pound, then probably the best amateur astronomer in the land. Molyneux commissioned from the famous instrument-maker George Graham a specially constructed telescope over 24ft in length,[11] and late in 1725 this was fixed with great care to a chimney stack within Molyneux's house, extending from the roof to near the floor. But whereas Hooke's object glass was rigidly fixed in direction and his telescope had no conventional tube, Molyneux's telescope had a tube pivotted at the top so that it could be moved very slightly to each side of the zenith in a north-south direction. As γ Draconis was observed passing overhead, the observer pointed the cross-hairs of the telescope directly at the star so that the star was in the middle of the field of view; the position of the star was then determined from the amount by which the telescope had been moved from the vertical, this being measured by means of a plumb-bob and graduated arc. This difference apart, Molyneux and Bradley were following Hooke in employing a telescope designed to observe one star and for one purpose only.

Not only for one purpose, but in only one coordinate; and in that coordinate, annual parallax should cause this particular star to reach an extreme southerly position each year in mid-December. They therefore worked hard in late November and early December 1725 to have the telescope erected and their checks on its accuracy completed by that time; this they did, and they then prepared for a long wait before anything of interest happened, for theory predicted that the star would be almost stationary for some weeks, after which it would begin to move steadily northwards, to reach its other extreme position in the middle of June.[12]

The delicacy of the mounting of the telescope was remarkable – the presence of three men standing perfectly still generated enough heat and air-movement to disturb the plumb-line that defined the vertical – and it is not surprising that on 21 December, some three days after the star should have reached its extreme southerly position, Bradley thought to observe the star again and check the performance of the telescope. To his amazement, the star passed more southerly than before, and this continued in the following weeks, until by March γ Draconis was no less than 20″ south of its December position, although by this time it ought, by annual parallax, to have been moving at maximum speed to the north. The star then stopped and soon began to return towards the north, passing through its December position in June, and reaching an extreme northerly position in September.

Their first thought – once it was clear that what was happening was no instrumental error – was that the axis of the Earth itself was changing direction, in which case the vertical plumbline by which Molyneux's telescope was set was also changing direction. The observed changes would then be, not in the star, but in the terrestrial coordinate system by which the star's position was being measured. Fortunately another star, on the opposite side of the North Pole from γ Draconis, also passed overhead, and although this star was not bright enough to be observed in the daytime, it could be followed sufficiently to show that its changes did not match those of γ Draconis in the simple pattern required by their hypothesis.

At the same time the two observers were considering an alternative explanation. They had followed Hooke in observing stars vertically overhead because such observations were free from the complication of atmospheric refraction. But this procedure was based on the belief that the atmosphere is a strictly spherical envelope surrounding the Earth, so that light from stars vertically overhead meets the atmosphere at right angles and is not bent by it. But suppose the atmosphere was not a spherical envelope. Suppose instead that it was distorted into some sort of elongated shape by the motion of the Earth through a resisting medium. In that case, light from a star vertically overhead would not necessarily meet the atmosphere at right angles; instead, it would be bent by its passage through the atmosphere, and by amounts that varied with the annual cycle of the Earth's movement around the Sun. This ingenious hypothesis encouraged the observers to look for certain particular patterns in the movements they were observing, but without success – the extraordinary observations of 1726 were so far inexplicable.

The immediate problem was to determine with confidence the *pattern* of behaviour of stars passing near the zenith at London, and for this more stars must be observed, which in turn required a telescope covering a wider band of sky than did Molyneux's. Bradley decided to commission one for himself from Graham, similar in principle, but of half the length, and giving access to stars many times further from the zenith.[13] He was allowed to install this in the home of Mrs Pound, one end of the telescope being at roof level and the other in the coal cellar, and holes being cut in the intervening floors. It was placed there in August 1727, and it was just as well that Bradley was no longer dependent on his friend, for Molyneux was to die early the following year.

Bradley was now able to observe no fewer than two hundred of the stars catalogued by Flamsteed, and of these he did actually observe some dozens with care. By about the end of 1727 or the beginning of 1728 the true pattern was clear to him: stars reached their extreme positions when they passed overhead at six o'clock, morning or evening; and they moved southward while they passed in the day, and northward while they passed in the night. The problem was to discover why.

The answer, we are told,[14] eventually came to him when he was on the River Thames on a pleasure boat, presumably in the summer of 1728. On the mast was a vane which showed which way the wind was blowing, and Bradley noticed that this altered direction every time the boat turned about. He commented on this to some sailors, who said that this always happened, and that it was due not to the changes in the wind but to the changes in the direction of the boat. (In the same way, if rain is falling on a calm day, when we stand in the street we hold an umbrella vertically over our heads; but if we begin to walk, we hold the umbrella in front of us because the rain *appears* to be beating in our faces.) Now, half-a-century earlier, the Danish astronomer Ole Römer had argued that light has a finite velocity in order to explain why eclipses of Jupiter's satellites were seen sometimes earlier and sometimes later than expected: when they were seen ahead of schedule, it was because the planets were so positioned that the light bearing the information of the eclipse from Jupiter to Earth had a shorter distance than usual to travel.[15] But though the speed of light was finite it was uniquely great, taking only a few minutes to cover the length of an 'astronomical unit' – the ninety or so million miles from the Sun to the Earth; and apparently it had never occurred to anyone that the speed of the Earth in orbit was in any sense comparable, and that the movement of the observer on Earth affects the direction of the light reaching him. For just as the wind direction as indicated by the vane on Bradley's boat seemed to alter as the result of the boat's altering direction, so the position of a star in the sky – that is, the direction from which the starlight reaches us – seems to alter as the Earth alters direction in its orbit around the Sun. The amount is small, for the Earth travels at only about one ten-thousandth of the speed of light; but as a result of this so-called 'aberration of light', stellar observations may be in error by as much as twenty seconds of arc.

Bradley had set out to measure annual parallax and so provide the expected observational proof of the motion of the Earth; he had found another proof, quite unexpected and more far-reaching in its consequences – so far-reaching that his published paper is arguably the most important in astronomy of the eighteenth century.

First, his discovery of aberration, combined with his subsequent discovery that the Earth's axis does wobble (or 'nutate') as a result of the attraction of the Moon on the non-spherical Earth, ushered in the era of exact positional astronomy. All previous observations had been made in ignorance of the errors introduced by these unsuspected complications, with the result that one observed position of a star might differ by over forty seconds of arc from the observed position of the same star six months later. Only when astronomers had been put on their guard by Bradley, and knew they must correct their observations to allow for aberration and nutation, could precision astronomy of position (and of changes in position, or 'proper motions') begin. Bradley later became Astronomer Royal at Greenwich and carried out a programme of exact and careful observations; and when Bessel later corrected these to allow for aberration and the like, the title he gave to his volume of Bradley's observations was no less than *Fundamentals of astronomy*.[16]

Second, Bradley's discovery enabled the approximate scale of the distances of the stars to be established for the first time. As we shall see (Section B, chap. 2), by assuming that the stars appear fainter only because they are more distant, Newton had placed the bright star Sirius at around one million astronomical units; but owing to a succession of accidents Newton's results were not published until 1728, only a few months before Bradley's paper. Newton's estimate depended fundamentally on the (highly questionable) assumption that all stars are physically similar, and on the basis of this assumption he derived an *actual* distance for Sirius. Bradley, by contrast, failed to find the annual parallax for which he was looking, and he considered that this must be because γ Draconis was so distant that the (double) parallax was too small for his telescope and so less than one second of arc. A simple calculation showed that the distance of the star must then be *at least* 400,000 astronomical units.[17]

The astronomical world was therefore, within a few months, offered two astonishing insights into the distances of the stars: Bradley's, that the *minimum* distance of γ Draconis was around 400,000 astronomical units; and Newton's, that the *actual* distance of the bright star Sirius was about one million astronomical units *provided* that star was physically the equal of the Sun. The minimum distance of γ Draconis derived by Bradley from direct observation therefore harmonised well with the actual distance of Sirius derived from an uncertain hypothesis, and together they provided a convincing demonstration that the nearest stars are some hundreds of thousands of astronomical units – a small number of light-years – from us. Bradley's paper helped establish the scale of distances in the stellar universe.

There were yet more results to be derived from Bradley's discovery of aberration. Previous estimates of the time light takes to reach us from the Sun had been based on studies of eclipses of Jupiter's satellites and had ranged from seven to eleven minutes. By a simple arithmetical calculation based on observations of aberration, Bradley was able to give the figure as $8^m\ 12^s$, within eight seconds of the modern value.[18] And because the different stars he had examined had all been affected by aberration to the same extent, he concluded that light from all these stars arrived at Earth travelling at the same speed – irrespective of whether the light had come from a near star or a distant one.[19] This was proof that light from the stars does not slow down in the course of its journey. Furthermore, since Bradley's figure of $8^m\ 12^s$ was roughly the same as the figures derived from eclipses of Jupiter's satellites – that is, since the figure derived from studies of direct starlight was roughly the same as figures derived from studies of reflected starlight – it followed that light travelled at much the same speed whether it was direct or reflected.[20]

It would be difficult to find any paper in the history of astronomy that had so many important results. Bradley had discovered and removed a major uncertainty in the measurement of stellar positions, and so introduced the era of precision astronomy; he had helped establish the order of distances in the stellar universe; he had found to within a few seconds the time light takes to reach us from the Sun; he had shown that light from distant stars reaches us at the same speed as light from near stars, and that reflected star-light travels at the same speed as direct light; and he had succeeded, in an unexpected manner, in fulfilling his original intention to prove the motion of the Earth.

REFERENCES

1. Page 9 of Robert Hooke, *An attempt to prove the motion of the Earth* (London, 1674), being a Cutlerian Lecture given at Gresham College in 1670. Hooke tells us in the preface that he planned to publish the lecture when it was first given "but was diverted by the advice of some Friends to stay the repeating the Observation, rather then publish it upon the Experience of one Year only". However, owing to sickness (and the accident to the object glass, see ref. 6), he had made no further progress by 1674 and "I do rather hast it out now, though imperfect, then detain it for a better compleating, hoping it may be at least a Hint to others to prosecute and compleat the Observation". As we shall see, his hopes were fulfilled.

    On p. 10 Hooke describes how he first discussed the problem at the Royal Society. Apparently this was on 20 June 1666: "He [Hooke] undertook to make observations of the parallax of the earth's orb to seconds; as also to make observations with long telescopes without the use of a tube" (T. Birch, *The history of the Royal Society* (London, 1756-57), ii, 98; *cf.* p. 109).
2. P. 8.
3. P. 10.
4. Pp. 22-23.
5. P. 24.
6. *Ibid.*
7. P. 25.
8. We have already seen (ref. 1) that Hooke's friends tried to disuade him from publication on the grounds that it was premature. From the Continent Huygens wrote that the observations were "of great Consequence; but they must be continued, to see, whether in the Course of one or more Years the *Parallaxes* do regularly answer to the Annual Motion of the Earth", and meanwhile he was planning observations from a vault 28 fathoms deep (*Philosophical transactions,* ix (1674), 90).
9. C. A. F. Peters, "Recherches sur la parallaxe des étoiles fixes", *Mémoires de l'Académie des Sciences de Saint-Pétersbourg,* sér. 6, Sciences mathématiques et physiques, v (1853), 1-65, pp. 6-7.
10. S. P. Rigaud (ed.), *Miscellaneous works and correspondence of the Rev. James Bradley* (Oxford, 1832), 93.
11. *Ibid.,* 93-115.
12. The story that follows is set out by Bradley in his great paper, "An account of a new-discovered motion of the fixed stars", *Philosophical transactions,* xxxv (1727-28), 637-61.

13. Happily, the instrument survives at the Old Royal Observatory, Greenwich (OM/Z.1). It is described and illustrated by Derek Howse, *Greenwich Observatory*, iii: *The buildings and instruments* (London, 1975), 60-64 and Figs 54-57.
14. T. Thomson, *A history of the Royal Society* (London, 1812), 346.
15. Römer's work is discussed in *Roemer et la vitesse de la lumière*, ed. by R. Taton (Paris, 1978).
16. F. W. Bessel, *Fundamenta astronomiae pro anno 1755* (Königsberg, 1818).
17. Bradley, "An account...", 660.
18. *Ibid.*, 653.
19. *Ibid.*, 654.
20. *Ibid.*, 653.

4. Goodricke, Pigott and the Quest for Variable Stars

By the 1770s the study of variable stars had been at a virtual standstill for several decades. There had been no repetition of the brilliant new stars of 1572 and 1604; and it was a century since Bullialdus had shown that Mira Ceti, the "wonderful star" in the Whale that had appeared and then disappeared and reappeared again, reached its maximum brightness every eleven months, and that, whatever the differences between one maximum and the next, to this extent at least its behaviour was lawlike and predictable.[1] Bullialdus had suggested that Mira may be a rotating star with dark patches, and such an explanation seemed plausible to most astronomers since the Sun itself was rotating; if the patches were thought to vary as do sunspots, then such an explanation could readily accommodate almost all variables except perhaps the most brilliant of novae, since any *periodic* component in the variation could be ascribed to the rotation and any *non-periodic* changes to alterations in the dark patches. Meanwhile, once it had been accepted in the late-seventeenth century that stars might change, alleged variable stars were proposed by the dozen, usually by observers who were deceived by changes in the seeing conditions or by erroneous entries in earlier star catalogues.[2] In 1715 Edmond Halley published a short paper[3] in *Philosophical transactions* listing the reliably-established "new" stars, six in number: the novae of 1572 and 1604, Mira Ceti, and three stars in Cygnus first noticed in 1600, 1670 and 1686 respectively. He should perhaps have added Algol (β Persei), seen in 1667 and 1670 as fourth magnitude instead of its usual second magnitude, but otherwise Halley could be said to have sifted out the indisputable variable stars from a welter of less convincing claims.

Early in 1780 Nathaniel Pigott, a gentleman and surveyor and an accomplished amateur astronomer, moved his home from Wales to York in the north of England where he expected to remain for the rest of his life.[4] As an astronomer his interests were conventional: comets, eclipses, transits of Venus and Mercury and the like. But as a surveyor he was exceptionally skilled in the application of precise astronomical measurements to the establishment of geographical positions, so much so that he had once spent five months on a survey of the positions of the principal towns in Belgium at the request of the Government, but at his own expense: "such reembursement to one in my position is never thought of."[5] In York Pigott set to work to construct "a magnificent observatory, built upon the model of that at Greenwich—stone foundation, stone solid Pillars, Slits in every direction &C",[6] with a transit instrument by Sisson and a quadrant by Bird; and these he used in the early 1780s to make frequent observations of zenith distances with a view to establishing the latitude of York.

Pigott's eldest surviving child, Edward (1753–1825), shared many of his father's interests, although father and son do not seem to have enjoyed a close relationship.[7] Like his father, Edward was highly competent in the use of astronomy in geodesy,[8] but he was otherwise much more enterprising in his astronomical pursuits. His interest in proper motions of stars is a good example of this.[9] In 1778 Edward had spent some weeks in Bath where he several times visited William Herschel,[10] then still a professional musician, and viewed with his "exceeding good Reflector", and the two men remained respected friends. Unlike Herschel, Edward Pigott possessed at York fixed instruments with which he could determine stellar positions with precision and so was able to investigate proper motions. On 16 February 1783 he wrote to Herschel mentioning that β Virginis seemed to have a proper motion in R.A. of more than one second of arc per annum, but only a third of that amount in relation to a faint star in the same constellation,

> which makes me suspect this Star, tho so small, to have also a proper motion; that some Stars of the 6th Mag^{de} are thus affected, I have much reason to believe, tho my computations on them are not yet compleated; and those are the Stars which I conjecture to be the nearest to our Earth; for if the distance of the Stars is according to their brightness, the motion of these small ones must be inconceivably rapid to be perceived by us, and infinitely more so than Sirius Arcturus & Procyon &c which is not probable.—how far the discovery of proper motions in small stars may induce Astronomers to choose them for finding a parallax I dare not venture to decide, but your opinion would give me particular pleasure.[11]

A more perceptive comment from the period, or one more critical to Herschel's own investigations, would be hard to find.

During the early autumn of 1781, the Pigotts had been hard at work positioning their fixed instruments.[12] Edward Pigott's first transit observation is dated 21 September,[13] and by the end of October he was in a position to resume his theoretical researches. But whereas in previous years he had shown little interest in variable stars, now he combed the literature for information on them and this he copied into the blank pages at the back of his old *Astronomical journal*.[14] He focussed his attention on the head and neck of the Swan (Cygnus), site of the 'novae' of 1600, 1670 and 1686, and he carefully copied a drawing by Gottfried Kirch showing the location of these stars.[15] On 28, 30 and 31 October he observed the transit of the nova of 1600, and on 3 November he "examined the Stars in the Swans head & neck for near an hour, with one of Dollond's *Opera Glass's* comparing them to Kirch's draught of Cygni".[16] The next two *Journal* entries, for 4 and 7 November, tell a similar story, and the next dated entry, for 14 November, announces the discovery of a comet in Cygnus close to the site of the novae.

A stone's throw from the Pigott home in York lived the Goodricke family, including the seventeen-year-old John, soon to become heir to his grandfather, Sir John Goodricke. John had been a deaf-mute from infancy and at the age of eight had been sent to a special school in Edinburgh. In 1778 he was able to enter the famous Dissenting school, Warrington Academy, where he evidently acquired a good grounding in mathematics.[17] On 16 November 1781, John Goodricke made the first entry in his *Journal of astronomical observations*: "Mr E. Pigott told me that at 9 o'clock P.M. yesterday he discovered a Comet."[18]

Pigott and Goodricke were to form an astronomical alliance that began virtually as that of instructor and pupil but soon evolved into a close partnership.

Pigott was already an experienced observer, technically accomplished, possessing fine instruments and with a network of scientific contacts. Yet, isolated in provincial York, and seemingly unsympathetic to his father, he yearned for companionship in astronomy and eagerly welcomed this new recruit. Even so, in 1783 he told Herschel "there is not a soul here to converse with" on astronomy[19]—for the deaf-mute Goodricke was unable to converse with anyone. No doubt these physical handicaps also reduced the possibilities for the two friends to work side-by-side at the telescope in the night darkness, although there are numerous occasions when Goodricke did visit his neighbour's well-equipped observatory. Certainly the difficulty of communication reduced the likelihood of any friction between them. In their published papers, and more revealingly in their private journals, each gives credit unstintingly to the other, and there is only one trace of a clash between them: on 31 July 1783 they had what Goodricke afterwards described as "too warm a dispute" as to whether one of the Cygnus novae had been visible the previous year,[20] but the next day Goodricke noted that Pigott had justice on his side and that the record on the previous page ought to be "blotted out of the Journal". It is amusing to picture this heated argument, presumably conducted through pencilled notes of ever-increasing bluntness!

Goodricke, then, in November 1781 acquired a mentor who was the proud discoverer of a comet, but who had chanced upon it in pursuance of his newfound passion for the study of novae and variable stars. Naturally the comet preoccupied the two acquaintances in the coming weeks, but early in December Pigott received a letter[21] bringing the latest data on Herschel's supposed comet (actually Uranus), and for a while many of their observations were of this. At the end of April, Goodricke acquired at last a telescope worthy of the name:

> All my observations on Herschells Comet & on that of Novr 1781 were observed only with a small perspective Glass magn. ab^t 10 or 12 times & their R.A. determined from their relative situation with respect to the stars ab^t them by the help of my Globe but having now just got a larger telescope of 2 feet & a half [an achromatic by Dollond] with two tubes mag. 80 & 50 times, I put cross wires in both these tubes in order to observe the Comet with greater accuracy. . . .[22]

In July Goodricke acquired a clock, which he systematically compared with Pigott's excellent timepiece, each clock being read when the chimes of York Minster struck.[23] Since Goodricke was deaf, he relied on the help of "some faithful person" for this. Later the accuracy of timing became so important that they had to allow for the difference due to Pigott's greater distance from the Minster.[24]

On 19 June Pigott at last resumed his study of the variable stars in Cygnus, examining the area with the same instrument as before. Perhaps it was next day that he gave Goodricke a copy of his sketch plan taken from Kirch[25] (it is repeated in Goodricke's *Journal* a few days later), for that evening Goodricke wrote: "Having a mind to observe the variable stars in Cygnus—Ab^t Midnight I looked for all the stars between γ & η Cygni in order to find the Nova of 1600."[26] Both men were now embarked on the researches that were to earn them immortality.

The same evening, a few hundred yards away, Pigott was also examining Cygnus. The following night Pigott made three separate searches of the area, and a fourth on 22 June; Goodricke observed Cygnus on 22 and 25 June. On 2, 4 and 10 July, Pigott searched without success for the nova of 1604, before returning to Cygnus on 24, 25 and 27 August and throughout September and October. Meanwhile Goodricke observed Mira decline from second to sixth magnitude in the course of five observations spread over the period from 9 August to 29 October. On 23 October, Pigott directed his attention to Algol and confirmed "this star is variable". He examined Algol again on 24 and 29 October and 1 November; on 29 October and 1 and 12 November he observed Mira.

At the beginning of November, Goodricke drew up for himself a summary of "Variable stars or those which are thought to be so",[27] beginning with Algol but otherwise leaning heavily on the mistaken assumption that Bayer at the beginning of the seventeenth century had assigned Greek letters to the stars of a constellation strictly in accordance with their descending order of brightness in his day. There is, however, no doubt of the confidence of the eighteen-year-old Goodricke in his ability to advance knowledge of variable stars. For some constellations he lists his own sequences of brightness; Algol, he noted, was now of second magnitude. On 4 November he observed Mira and checked his sequences for Hercules and Draco. On 7 November he re-observed Hercules, and found Algol "the same as before". But when he looked at Algol on 12 November he was astonished to find it was now of the fourth magnitude:

> This night I looked at β Persei & was much surprized to find its brightness altered—It now appears of ab^t the 4th magn^d. It was hardly distinguishable from ρ Persei.... I observed it diligently for ab^t an hour & upwards—I hardly believed that it changed its brightness, because I never heard of any star varying so quick in its brightness—I thought it might perhaps be owing to an optical illusion or defect in my eyes or bad air but the sequel will shew that its change is true & that I was not mistaken.[28]

We can be sure that he told Pigott, for both men observed Algol the following night, but they found it back to second magnitude. Goodricke looked again on 15 and 26 November, and Pigott observed Algol's transit on 12 and 15 December, but with the star apparently restored to stability their *Journal* entries are devoted to other topics—to other variables, to Uranus, and to an occultation of Spica by the Moon.

Then, on 28 December, both men saw Algol again make the dramatic drop in brightness. Pigott described the changed in his *Journal*:

> Algol ... this star being discovered to b[e] Variable, by Montanary & Maraldi, I began, on the 23^d of October, to examine it ... & have continued ever since when I was not otherwise engaged—on the 12th of Novem^r M^r J Goodricke told me he had found it, the preceding night, only of the 3^d Mag^{de} ... every night since I have been constantly very attentive in examining it, and saw no alteration, not even on the 23^d–24th & 26th but this night found it of about the 3^d or 4th Mag^{de} having the same brightness and colour as δ Persei ... at about 6^h P.M. it had become brighter ... at

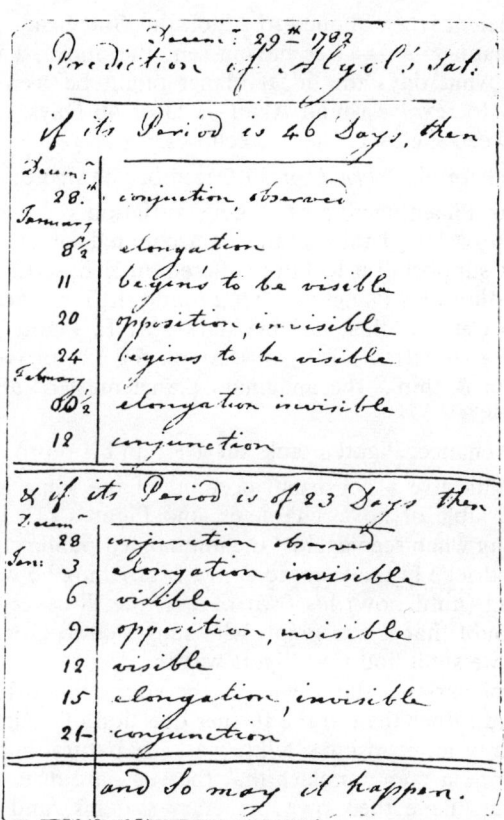

FIG. 1. Evidence that Edward Pigott examined the implications of a supposed satellite of Algol on the day following the York astronomers' second observation of the star's reduction in light. On this scrap of paper in Pigott's hand, he calculates when the satellite would be on the further side of the star and possibly visible from Earth, on the assumption that in the 46 days since Goodricke's first observation the satellite had completed either 1 or 2 orbits. In fact the satellite had completed no fewer than 15 orbits. (From the Edward Pigott MSS, North Yorkshire County Council.)

$\frac{1}{2}$ past 8^h P.M. it had recovered its former brightness... and at 12^h I thought it rather more brilliant than I had ever seen it.[29]

Next day Pigott sent Goodricke a note asking for a copy of the observation of 12 November, no doubt to establish exactly the time of the previous minimum and the interval between them. In the first week of January he made a record of the suggestion that accompanied the request:

... the opinion I suggested was, that the alteration of Algol's brightness, was maybe occasioned, by a Planet, of about half his size, revolving round him, and therefore does sometimes eclipse him partially; which accounts also for the unequal time (if any) of the duration and Periods of this phenomenon—the various Systems attempting to account for the changes in all the other variable Stars, as o Ceti [Mira] χ Cygni [of 1686] &c &c,

are very deficient when applied to Algol, this Star being variable in a very particular manner————after having sent the above, I drew up a table, shewing on what days this ideal Planet might be seen from the Earth, supposing it to revolve round Algol in 23 or 46 Days, this last being the interval between the two observed eclipses.[30]

Goodricke agreed. In his *Journal* for 30 December he wrote:

The Singular Phaenomenon of Algol's variation on the 28th inst. & on the 12th of Novr last, I think, can't be accounted for in any other manner than that of supposing it to have suffered an Eclipse (if I may say so) by the interposition of a Planet revolving round it. This variation is evidently difft from o Ceti & other variable stars—Mr E. Pigott having sent me a note to desire an extract of my observn on the 12th of Novr last, gives the same opinion & thinks the imaginary planet must be abt half the size of Algol at least.[31]

By a remarkable chance, Pigott's table survives, and is reproduced as Figure 1.

These documents give a consistent account of the steps in the recognition of Algol as a variable of novel character, and Pigott's private statements are entirely convincing when seen against the unqualified public credit he invariably accorded to Goodricke in the years to come. If anything he was over-generous, totally concealing (until now) his own part in the discovery. For just as we can no longer doubt that it was Pigott who suggested to Goodricke that Algol was eclipsed, so we shall find that Pigott was for decades to speculate over the physical nature of variable stars, whereas the young Goodricke's interests lay in the phenomena rather than in the further question of their causes.

The suggestion of an eclipsing satellite was very natural in the context of the time. Just as it was a commonplace that the stars are distant suns, so it was accepted that stars have their own planetary systems; and if a *large* planet circled round a star in a plane that chanced to pass through the solar system, then perceptible eclipses would occur. But the hypothesis was vulnerable: for it to be sustained, at least in its simplest form, observations would have to confirm not only that the eclipses occurred at regular intervals but also that, unlike Mira, Algol's brightnesses at successive maxima and minima were consistent and the rises and falls symmetric. Even so, Algol would then have to be singled out as a variable different in nature from most if not all the others known; for the rest the "dark patches" explanation would continue.

Naturally both men now kept a close watch on Algol, and both set aside pages at the back of their observing journals for the purpose. On 14 January Goodricke estimated that the "supposed eclipse" took seven hours and commented: "If the Period of Algol's variation is regular, this will prove that it has a planet revolving round him."[32] By 31 January he thought the period to be about 17 days; on 6 February he speaks again of the "supposed eclipse". By mid-April the evidence had accumulated that the true period was astonishingly short: some $2^d 20^h 45^m$. Publication was delayed while Pigott tracked down the references to changes a century earlier, by Montanari and Maraldi, though these were to prove of little use.[33] But meanwhile Pigott, who had been notably reticent about Algol when writing to Herschel on 16 February,[34] now wrote (or spoke) to the Astronomer Royal, Nevil Maskelyne, telling him in confidence of the discovery of the period and of the proposed explanation, and asking him to pass word to Herschel.[35]

Maskelyne's letter reached Herschel on 27 April, and he at once began to keep watch on the star.[36] Meanwhile, London scientific circles were abuzz with the news. On 30 April Sir Joseph Banks, President of the Royal Society, wrote to Herschel (with no request for confidentiality) to say that "Mr Pigot [!] is said to have made a most interesting discovery, that one of the fixd stars has a planet revolving round it so exactly in the plane of our orbit as to eclipse the star partially each revolution[;] the period is nearly 3 days & the diminution of Light can be foretold exactly, as well as it[s] return. The star is algol ..."; later the same day he sent word again to Herschel asking if he might himself come down to Datchet that evening to share in the observations. He did so, but without seeing the drop in brightness; on 2 May he wrote again to say that at the Royal Society he had learned that the "occultation" had happened around midnight, and that the discovery was due, not to Pigott, but to "a deaf & dumb man, grandson of Sr John Goodrick who has for some years amused himself with astronomy".[37]

Meanwhile Herschel was observing Algol as often as possible. As he explained to Pigott later that month:

> Having occasion to go to London on Thursday [8 May] I drew up my observations on Algol that morning & put them in my pocket, with an intention to shew them to my astronomical friends.
>
> Not arriving in Town till past 4 o'Clock I went immediately to the Crown and Anchor, where I was directly asked whether I had seen this curious phenomenon; to which I answered that I had—and brought an account of it with me. Several gentlemen expressed a desire of hearing it, and S^r J Banks read it after dinner. The paper thus being in his hands, it was thought proper to read it also in the evening. Thus you see Sir that the paper was read without the least thoughts, on my side of any paper of Mr Goodrike's. It was an affair publikly known so that I might naturally suppose some paper of yours or Mr Goodriks had been communicated upon the subject.[38]

In fact Goodricke did not dispatch his paper until 12 May, and it was read to the Royal Society on the 15th.[39] Pigott had left York a month or so earlier,[40] and it may be that his absence was partially responsible for the cautious stance Goodricke adopted over the physical causes possibly at work:

> If it were not perhaps too early to hazard even a conjecture on the cause of this variation, I should imagine it could hardly be accounted for otherwise than either by the interposition of a large body revolving around Algol, or some kind of motion of its own, whereby part of its body, covered with spots or such like matter, is periodically turned towards the earth. But the intention of this paper is to communicate facts, not conjectures....[41]

The *Dictionary of scientific biography* suppresses the second "alternative cause" when quoting this passage,[42] and so presents Goodricke as totally committed to the eclipse theory, which spectroscopic examination a century later showed to be correct. But it could be argued that of the principals involved Goodricke was the least to be thus committed. Pigott seems to have left Maskelyne and the rest of the London circle in little doubt that the variations were indeed due to eclipses; while Herschel in his record of observations of 3 May writes of "the star's occultation" and of its being "much eclipsed", and in the discussion read for him by Banks he hails the "present observations [which] will probably furnish us with the strongest arguments and facts to verify former conjectures, of a plurality of solar and planetary systems".[43]

It is not easy to determine exactly when the eclipse hypothesis was abandoned. Herschel wrote to Lalande on 23 May to report, among other things, the variations in Algol, but he gives no hint as to a possible cause;[44] however, this may be no more than his usual veneration for the unvarnished observational facts—not for him the modern philosopher's dictum that "all facts are theory-laden". Goodricke, in a second Algol paper dated 8 December the same year,[45] uses his own subsequent observations of Algol together with one of Flamsteed from 1696 to amend the period to $2^d 20^h 48^m 56^s$, within seconds of the modern value, but he has nothing to say about causes; true, in the manuscript he does alter "periods" to "revolutions",[46] and it is probable that this refers to revolutions of the satellite, but it could also mean rotations of Algol itself. Meanwhile, in private he was noting apparent irregularities not readily compatible with the eclipse hypothesis. On 15 August 1783 he comments: "I am almost sure that Algol was rather varied from its usual brightness."[47] On 20 October 1784 "Algol was not so much diminished as it generally is".[48] Given the sensitivity of the observations to changes in seeing conditions, it would be surprising if suspicions of irregularities did not accumulate. But the final blow to the eclipse hypothesis was probably the discovery (to which we now come) by "the three York astronomers"[49] in the autumn of 1784 of more variable stars with short periods like Algol—but stars some at least of which could not be explained on the eclipse hypothesis.

Much of Goodricke's *Journal* for 1783 is naturally taken up with Algol. He several times observes Mira and χ Cygni, and there are references to an "excellent" equatorial by Ramsden he acquired in May,[50] to an eclipse of the Moon in September, a comet discovered by Pigott in November, and other matters of natural interest to an astronomer. But with the completion of the (first) Algol paper in May, Goodricke began to broaden his search for variables. He several times compared the order of brightness of the stars in Hercules with his corresponding list of the previous November, but without any firm conclusion. On 21 October he began to suspect changes in α Cassiopeiae, a star which he often used for comparison with Algol; but he does not seem to have followed this up or to have discussed it with Pigott. (In July 1787, after Goodricke was dead, Pigott independently came to the same conclusion, noting "I am now of opinion that the α varies in a few hours",[51] and setting aside a page at the back of his *Journal* for his observations of the three stars α, β and γ; but he had the greatest difficulty in deciding which of the three varied, and scraps of related observations survive from as late as 1812 and 1815, long after Pigott had lost touch with his fellow astronomers. Curiously, at the beginning of the twentieth century[52] it was accepted that α Cassiopeiae varied by half a magnitude, but α is now believed to be constant while γ is known to be subject to considerable variations.)

During the same year of 1783 Pigott likewise gave the bulk of his *Journal* to the study of variables, but we find him ranging far more widely than Goodricke in his search for changes. On 10 December he saw for the first time the variable star in Hydra (R Hydrae) studied by Maraldi at the beginning of the century and believed to have a period of some two years. This Mira-type variable was to become a firm favourite with Pigott; in 1786 he published a much-improved estimate[53] of 494 days for its period, and he continued to observe the star until at least 1815.[54]

In the spring of 1784, Goodricke, who was then nineteen and already a Copley medallist of the Royal Society, spent some weeks in London "in order to improve himself in the different branches of astronomy",[55] and he was kindly received by Maskelyne.[56] As an observer he had an uneventful year until August, when he compared the constellations of Lyra, Capricorn and Aquarius with their representations in Flamsteed's *Atlas*. On 10 September "I thought β Lyrae was much less than usual. It was much less than γ Lyrae & was less than a star of the 4th magnitude . . .".[57] He observed the star carefully in the following days, and by 30 September had assigned it a period of either 13 days or 6 days 10 hours. In fact β Lyrae is a double star in which the brighter and fainter companions are thought to be very close together and ellipsoidal in shape as a result of tidal distortion; each is eclipsed once in 12·91 days, with a large drop in brightness alternating with a somewhat smaller drop. In a paper[58] dated 10 January 1785 and read to the Royal Society on the 27th, Goodricke shows an excellent understanding of the light cycle which he analyzes as follows: two days at mag. 3, followed by a reduction in brightness taking $1\frac{1}{4}$ days, then under one day at between mag. 4 and mag. 5, an increase taking two days, three days at mag. 3, a reduction taking one day, then under one day at slightly above mag. 4, then an increase taking $1\frac{3}{4}$ days—the whole taking 12d19h. It was not a cycle that could readily be accommodated to any simple eclipse theory, and Goodricke ascribes it to "a rotation on the star's axis, under a supposition that there are several large dark spots upon its body", as an afterthought giving himself further flexibility by adding: "and that its axis is inclined to the earth's orbit."[59]

In his paper Goodricke promises a further and more exact account, which he did not live to complete. Partial drafts survive among his manuscripts, and it is clear that he began to think that the minimum brightnesses of β Lyrae vary considerably and do not simply alternate as he had previously thought. His (unpublished) conclusion, which embodies his maturest thoughts on the mechanism at work, is this:

> . . . It appears to me now that the variation of β Lyrae should be arranged into four points only as those of η Antinoi & δ Cephei & consequently its period will be much less than that established in my first paper. This period as will hereafter appear is 6–0–0. Its changes, which seem always to be regulated by the degree of its diminution, which is different in different periods, are expressed as follows.
>
> 1. When the star goes down to between the fifth & fourth magnitude & its brightness nearly equal to that of ζ or κ Lyrae. They are thus
>   1. It is at its greatest brightness 2 days & twelve hours

2. It diminishes in 1ᴰ-8ʰ-
 3. It is at its least brightness 0ᴰ-21ʰ-
 4. It increases in 1ᴰ-12 hours
2. When it goes down to about the fourth magnitude & its brightness nearly equal to that of θ or ξ Herculis.
They are thus
 1. It is at its greatest brightness 3ᴰ-6ʰ
 2. It diminishes in   1 -0
 3. It is at its least brightness   0 -22
 4. It increases in   1 -12

When the star is diminished to a magnitude & brightness different from either of the above, which it sometimes does, its changes may easily be found by comparing the two sets together. I have already remarked in my first paper that the magnitude of the star at its greatest brightness is the third & that it is then sometimes equal to & sometimes greater than γ Lyrae, to which I may now add from some observations of a later date that it is sometimes less than γ Lyrae & of between the third & fourth magnitude. I have also noticed in the same paper that when the star is in that point, it is subject to some occasional variations, which I find is also the case sometimes at present.

Though it is above said that the star is unequally diminished in different periods & that as it should seem without any regard to stated times, yet I must here remark that even in this respect the star is generally subject to some regular Laws. This will be very apparent when all the observations contained in my first paper are examined, where it will be found that though the Diminution in the next following period was considerably different from the preceding, yet the same changes always occurred after one of the periods has elapsed or after an interval of 12ᵈ-0-0. I however find from my later observations that this circumstance did not always attend the star tho' upon the whole it seemed to happen oftener than the contrary. This circumstance, which is certainly very remarkable forasmuch as it does not appear in the other variable stars tho' I have remarked something of that kind in Algol & δ Cephei [on which see below] but it was not in a very considerable degree & as yet with no certainty of their happening at stated times, seems to be caused by an alteration of the Inclination of the stars axis to the Earth's orbit or which explains the matter better some kind of a libration & this alteration or libration must at least be performed constantly & regularly tho' sometimes subject to some deviations as above mentioned.[60]

For Pigott much of 1784 was devoted to his programme of "verifying all the stars suspected to be variable",[61] and on 10 September (the very day on which Goodricke noticed the variability of β Lyrae) he returned to θ Serpentis, said by Montanari to be variable. He had noted the previous year that it was fainter than η Antinoi (η Aquilae), but now their order was reversed: "... it seems that η Antin: has changed a little." By 20 September he had already given the variation a provisional period of 7 days, and in his paper[62] dated 5 December he refined this to 7ᵈ4ʰ38ᵐ. He also realised that the star takes much longer to decrease than to increase (it is in fact a Cepheid variable), and though this was incompatible with a simple eclipse hypothesis Pigott is ambivalent:

> Hitherto the opinion of the astronomers concerning the changes of Algol's light seem to be very unsettled; at least none are universally adopted, though various are the hypotheses to account for it; such, as supposing the star of some other than a spherical form, or a large body revolving round it, or with several dark spots or small bright ones on its surface, also giving an inclination to its axis, &c.; though most of these conjectures with regard to Algol be attended with difficulties, some of them combined do, I think, account for the variation of η Antinoi.[63]

A fourth variable of short period was now to fall into the net of the York astronomers. Whereas the other discoveries emerged almost naturally from their programme of observations, it seems to have been sheer familiarity with the sky that led Goodricke to suspect on 20 October "some variation amongst the stars ζ, ι, δ, ϵ, ξ Cephei".[64] By 23 October he wrote "I am now *almost* convinced that δ Cephei varies".[65] It was in fact the following June before Goodricke sent his results to the Royal Society for publication, along with his thoughts as to the physical explanation:

> What I have before mentioned, that the greatest brightness of δ Cephei does not seem always to be quite the same, is not peculiar to that star, but is also to be observed in the other variable ones. I have remarked in a late Paper that the greatest brightness of β Lyrae is subject to considerable alterations, and thought then that it might be owing to some fallacy of observation; but now I have reason to alter, in some measure, my opinion on this head. Even Algol does not seem to be always obscured in the same degree, being perceived to be sometimes a little brighter than ρ Persei, and sometimes less than it. These seeming irregularities however do not appear to affect the period, for if we compare the same precise phases together, it will be found still regular. This may, I suppose, be accounted for by a rotation of the star on its axis, having fixed spots that vary only in their size.[66]

To Herschel, meanwhile, Goodricke had written:

> I agree with you in thinking, that Algol's changes are occasioned by a rotation on its axis; I suppose you account for it under a supposition either that it is a spot of light & its axis inclined to the Earth's orbit: or that it is a much flattened body. The variations of β Lyrae & η Antinoi also appear to be occasioned by a rotation on their axes.[67]

Though he could not know it, Goodricke's life-work was almost complete. In 1785 entries in the body of his *Journal* become rare, and the bulk of his efforts went into repeated observations of the known variables, which he listed separately—δ Cephei, for example, he observed on one hundred nights in the first ten months of the year,[68] but he did not live to complete a second paper on the variable in which he would have given the interval from maximum to minimum as over five times that from minimum to maximum. On 30 March 1786 he observed β Lyrae for the twelfth occasion that month, and on 20 April he died, reputedly of an illness due to exposure to the night air.[69] He was 21 years of age and had been an FRS for just two weeks.

A few weeks earlier, Pigott had drawn to a temporary close his programme of observations of previously suspected variable stars, with a major survey in *Philosophical transactions* of all stars known or suspected of variability.[70] To progress in understanding the causes at work, he there argued, astronomers must first classify variables into three groups, those which are known to have appeared only once, those with long periods, and lastly a group containing the four short-period variables he and Goodricke had discovered: as to this last group, "the cause of their changes seem [*sic*] in general to be attributed to spots, and a rotation on their axis. This property of the fixed stars, though often suspected, was far from being evident till within these two years."[71]

At the end of 1785 Pigott noted in his transit book: "As we propose staying abroad for about two years, the instruments were all taken down, well cleaned, packed up & put in a dry Garret." On 16 March 1786 he sent from York to the Royal Society a paper on the determination of the latitude and longitude of York,[72] but he and his father were in Louvain for the transit of Mercury on 3 May,[73] and they doubtless had already left York when Goodricke died. "This worthy young man exists no more; he is not only regretted by many friends, but will prove a loss to astronomy, as the discoveries he so rapidly made evince."[74]

In fact the York observatory had almost exactly spanned the few short years of Goodricke's career as an astronomer, and Edward Pigott's help had been decisive. Goodricke's *Astronomical journal* and other manuscripts reached Pigott in November 1791 "being a present from the G: Family at York; knowing the value I set on every thing that belonged to my late most worthy and intimate friend". The *Journal*, he discovered, was "very similar to this of mine", as indeed it was, with the same page size and layout, the same headings and columns, the same style of index, and the same assembled lists of observations of variables of special interest. Probably only at that moment did Pigott realise how much Goodricke had depended on him for guidance: "my Friend departed this World April 1786, an event I shall ever lament."[75]

The Pigotts remained in Louvain until the middle of 1787, when they returned to England and spent the next years in and around London.[76] They returned to York at the beginning of January 1792, but in August that year Edward's beloved mother died,[77] and on 22 January 1793 he wrote: "Here concludes the Observations made in our Observatory in York: where for the future I do not mean to reside: besides my Father has parted with the lease of his House & consequently removed all his instruments." He remained in York for some months, but moved to Bath late in September and purchased a portable transit instrument by Sisson.[78] With this he observed systematically until June 1796, when entries in the *Transit book* cease. Otherwise the great majority of his observations in the middle and late 1790s are of variables, which were now among the objects most accessible to his limited instrumental resources; and although he concentrates on the familiar stars, his investigations were to be rewarded with two further notable successes.

Back in the summer of 1783, Corona Borealis had been among the constellations the brightness of whose stars Pigott had carefully recorded.[79] He had returned to the constellation several times later that summer, and again the following July, when a faint star "struck me as being considerably brighter".[80] In Bath on 16 May 1795 he examined the star once more, and was immediately convinced it had grown faint again—"in fact it is with the utmost difficulty that I can see it at all".[81] It was the irregular variable R Coronae Borealis, for which Pigott tentatively proposed a period of $10\frac{1}{2}$ months.

The second discovery was of a star that is never much above fifth magnitude, in the constellation Scutum. On 25 September 1795 he wrote: "I am greatly inclined to think this star is variable & that it has been increasing in brightness since at least a month."[82] Presumably Pigott kept notes which he never transferred into his *Journal*, but even so this detection suggests remarkable familiarity with the usual appearance of that part of the sky. To this star (R Scuti, a semi-regular of the RV Tauri type) he assigned a period of 63 days. The two discoveries were announced in *Philosophical transactions* for 1797.[83]

A feature of Pigott's journal for this period is the frequent speculations as to the physical causes of stellar variation. In 1788 he notes various features common to groups of variables, and wonders if χ Cygni and R Hydrae are moving "in an orbit of a narrow parabola" (!), and if so whether the changing distance would result in a significant change in the times taken by their light to reach us. He adds: "the transmission of light may be a perceptible equation for Algol."[84] In 1791 he again sought to explain variations in the periods of these two stars and of Mira as being "occasioned by the time light takes to pass over their different distances from us—this is supposing that they move rapidly in vast orbits".[85] Later the same year he toys with the idea of explaining variations through the mechanism of "a luminous fluid roling on the Star".[86] In 1795 he is back to "supposing them to move in an Elipse? this is giving them an immense rapid motion . . .".[87] Variables with periods of several months he accounts for by assuming "1. the period of the Star is that of its motion round a center or revolution— 2. one hemisphere of the Star is bright the other dark— 3. its rotation on its axis is equal to its revolution . . . but perhaps there is no difficulty, in the more simple supposition of a rotation on their axis with spots."[88]

That same year, Herschel published a paper in *Philosophical transactions* in which he provided a physical basis for his long-standing conviction that the Sun is habitable, by arguing that the Sun consists of a dark central body surrounded by a luminous atmosphere. The stars, he argued, are also habitable, for the variations show their nature is that of the Sun:

> The sun turns on its axis. So does the star Algol. So do the stars called β Lyrae, δ Cephei, η Antinoi, o Ceti, and many more; most probably all. From what other cause can we so probably account for their periodical changes? Again, our sun has spots on its surface. So has the star Algol; and so have the stars already named; and probably every star in the heavens. On our sun these spots are changeable. So they are on the star o Ceti. . . .[89]

Also in 1795 Herschel told Pigott that he had constructed "models that represented the diminution of the light of these stars",[90] and the following year he prefaced his "Second Catalogue of the Comparative Brightness of the

Stars" (catalogues in which he left to posterity a careful record of sequences of brightness in his day) with further "Remarks tending to establish the rotatory motion of the stars on their axes"; in these he allowed not only rotations but also "other movements, such as nutations or changes in the inclination of their axes; which, added to bodies much flattened by quick rotatory motions, or surrounded by rings like Saturn, will easily account for many new phaenomena that may then offer themselves to our extended views."[91]

After 1795 there are no further speculations as to the physical causes of variability in Pigott's *Journal*, in which the latest entries are dated December 1801, but other documents of this period show him strongly influenced by Herschel's papers and accepting his physical explanation of the structure of the stars. A handwritten list[92] in his copy of Flamsteed's *Atlas* is headed "*Periodical Rotations* of some of the Variable Stars" (italics supplied), and a lengthy list of "Memorandums" about the luminous fluid on the surface of the bodies of stars refers repeatedly to Herschel's paper (or papers).[93] Eventually, in 1805, Pigott published his last paper[94] on variable stars, consisting of a second report on R Scuti "exhibiting its proportional illuminated parts, and its irregularities of rotation", together with a public statement of his views as to the physical causes of variable stars. His theory was:

1st. That the body of the stars are dark and solid.
2d. Their real rotations on their axes are regular.
3d. That the surrounding medium is by times generating and absorbing its luminous particles in a manner nearly similar to what has been lately so ingeniously illustrated by the great investigator of the heavens, Dr. Herschel, with regard to the sun's atmosphere....[95]

Following Herschel's example in constructing working models of variable stars, "I have tried practically the effect of the above suppositions, by placing small white spots on a dark sphere, which being revolved round represented the various changes as nearly as could be expected",[96] and in diagrams he attempts to convey to the reader the visual effects of these models. Naturally Pigott has no difficulty in explaining all manner of changes with the aid of his luminous particles; and he further adds that Mira, Algol, and Herschel's variable α Herculis, must have been constant in Antiquity (since otherwise their changes must surely have been noticed), while other bright stars once variable now seemed to be constant.[97] No doubt, he continues, there are many "unenlightened" (dark) stars, perhaps collected together as are the enlightened stars of the Milky Way—in which case "they would appear as *dark spaces* in the heavens, similar to what has been observed in the Southern Hemisphere".[98]

Pigott promised one more paper on the subject, "probably my last",[99] but it was not to be. He had returned to the Continent following the Treaty of Amiens in 1802, only to be detained at Fontainbleau when hostilities broke out in 1803, and his 1805 paper is written partly in 1802 from Bath and partly in 1803 from Fontainbleau. Indeed, it seems that Pigott must have got word to Banks of the contribution he wished to forward from Fontainbleau, for on 30 January 1804 Banks wrote to Delambre:

> ... There is among the English detain'd in France a Mr. Edwd Pigot who has some astronomical knowledge. He has written a paper on that subject

which he wishes to be forwarded to the Royal Society. If you Sir could give him any assistance in forwarding it I shall be oblig'd to you.[100]

Banks later claimed, no doubt with justice, that Pigott's release in the summer of 1806 was due to the intercession of Banks's French friends.[101] Pigott's father had died in York in 1804 and the York instruments were now his son's property;[102] at a meeting of the First Class of the *Institut* in Paris on 26 May 1806 a letter from Pigott was read "pour prier la Classe de demander pour lui au Gouvernement la permission de retourner en Angleterre où ses instrumens d'astronomie et ses collections de botanique dépérissent en son absence", and by mid-July his passport had been returned to him.[103]

Back home in Bath, Pigott sent for his father's instruments,[104] but there is little sign that they were put to any effective use. He made observations of comets,[105] notably that of 1811; and his *Journal* and his copy of Flamsteed's *Atlas* are stuffed with loose scraps of paper covered with observations, some dated as late as 1815, of the variables he had been studying on and off for so many years. But his productive period was over. In 1817 he wrote to Herschel reminding him that it was nearly forty years since they first met and hoping they would soon meet again,[106] and it may have been at Herschel's suggestion that John Herschel invited him in 1821 to be a member of the future Royal Astronomical Society.[107] The opportunity came too late. "I am extremely sorry to say", Pigott replied, "that my health Capacity and Situation here oblige me to decline the very Polite & kind offer you do Propose . . .".[108]

History, as if to compensate John Goodricke for his physical handicaps and the brevity of his life, has been more than generous to his memory. His name is familiar to modern astronomers as that of the discoverer of the periodicity of Algol (and of those of β Lyrae and δ Cephei), and he is commonly said unequivocally to have advocated an eclipse theory for Algol. The time has come to do justice to Edward Pigott. It is now clear that it was he who introduced the boy Goodricke to the study of variable stars, and that under Pigott's leadership the two observers formed in effect a team for the nightly scrutiny of variables. By chance it fell to Goodricke to see Algol drop in brightness, but it was Pigott who in writing proposed an eclipse theory to his friend (who may, or may not, have already hit upon the idea) and Pigott who computed a table to show when the eclipsing body might be at maximum elongation and possibly accessible to observation. And it was Pigott who electrified the London scientists with his dramatic announcement of the eclipses.

In the $4\frac{1}{2}$ years of activity on the part of the York astronomers, Pigott found only one short-period variable, η Aquilae, to Goodricke's three. But Pigott was to discover two other important variables, R Coronae Borealis and R Scuti, and he deserved to have more to show for his many years of study of other possible variables, such as α, β and γ Cassiopeiae. Equally important was his major survey in 1786 of known and suspected variables, which proved a starting point for later investigations. Lastly, Pigott now emerges as a persistent if reckless theorist as to the causes of variation, his speculations beginning with the eclipse theory and passing through many vicissitudes, until the seemingly universal absence of regularity forced him to follow Herschel in adopting the quasi-traditional and infinitely flexible explanation of variations, as due to changes in the luminous atmosphere of a rotating dark body. As he remarked in "Memorandums" written about the turn of the century:

...it appears that *most all* these Variables attain in different periods different degrees of brightness both when at their Maximum and Minimum ... that they are subject to evident irregularities in their various Phazis, is also remarkable—I own that these two circumstances have ever surprised me much, not that some or most of them, but that not one of them should be free from perturbation seems to deviate from the wonderfull regularity which is apparent in the other celestial bodies.[109]

More was now known of variable stars; but their physical secrets they had preserved intact in the face of the most persistent and skilled investigation prior to Argelander's campaign half a century later, when astronomy was on the threshold of the spectroscopic era.

Acknowledgements

The author is greatly indebted to the Yorkshire Philosophical Society and the North Yorkshire County Council who, through the good offices of Professor M. M. Woolfson of the University of York, agreed to place the Goodricke and Pigott Archives on temporary loan in the Archives Centre of Churchill College, Cambridge.

REFERENCES

1. Ismael Bullialdus, *Ad astronomos monita duo* (Paris, 1667), Monitum primum: De stella mirabili quae in Collo Ceti conspicitur.
2. G. Montanari's letter to the Royal Society in 1670 claiming he had observed changes in a hundred stars (*Philosophical transactions*, vi (1671), 2202) made a deep impression.
3. [E. Halley], "A short history of the several new-stars that have appear'd within these 150 years ... ", *Philosophical transactions*, xxix (1714–16), 354–6.
4. On Nathaniel Pigott, see the articles in *Dictionary of national biography* (by Agnes Clerke) and *Dictionary of scientific biography* (by Zdeněk Kopal). The *DSB* article must be used with caution; in particular it overlooks the important paper by Sidney Melmore, "Nathaniel Pigott's observatory 1781–1793", *Annals of science*, ix (1953), 281–6. That Pigott moved to York in 1780 is established by statements in the letter of 9 June 1781 from Edward Pigott to William Herschel, Royal Astronomical Society Herschel MSS, W.1/13.P.27; indeed, Nathaniel Pigott wrote to Maskelyne from York on 23 May 1780 (copy in Nathaniel Pigott MSS, North Yorkshire County Council). Nathaniel Pigott tells Herschel of his intention to remain in York for life in his letter of 17 June 1782, Herschel MSS, W.1/13.P.43.
5. Melmore, *op. cit.*, 285–6.
6. Pigott to William Herschel, 17 June 1782; Herschel MSS, W.1/13.P.43. On 27 August 1777 Thomas Sisson sent Pigott a model of the well-equipped observatory of the noted amateur, Alexander Aubert (his accompanying letter is in the Royal Astronomical Society archives, Pigott 75).
7. On Edward Pigott, see the articles in *Dictionary of national biography* (by Agnes Clerke) and *Dictionary of scientific biography* (by Zdeněk Kopal).
8. Witness his elaborate article on "The latitude and longitude of York determined from a variety of astronomical observations", *Philosophical transactions*, lxxvi (1786), 409–25.
9. In the back of his *First astronomical journal* (Edward Pigott MSS, North Yorkshire County Council (hereafter: EP MSS)) he had already listed stellar positions as given by Tycho and Flamsteed in the hope of deriving "a particular motion in any of the fixt

stars", and in his letter to Herschel of 16 February 1783 he says "My time has also been much taken up in making observ^{ns} for determining the proper motion of several stars" (Herschel MSS, W.1/13.P.29).
10. Pigott, *First astronomical journal*, entries for 12 April 1778 *et seq*.
11. Original in Herschel MSS, W.1/13.P.29.
12. Edward Pigott to William Herschel, 9 June 1781; Herschel MSS, W.1/13.P.27. See also Pigott's *Transit book* (EP MSS).
13. See his *Transit book*, 23.
14. See Pigott's *First astronomical journal*. The period when he combed the literature is indicated by the parallels between these notes and the entries about the beginning of November 1781 in his main *Journal* (EP MSS).
15. The Kirch drawing he takes from *Miscellanea berolinensia*, i (1710), 210.
16. Pigott, *Journal*, 112.
17. "1778. John Goodricke, York. Son to Henry Goodricke, Esq., M.P. He lost his hearing by a fever when an infant, and was consequently dumb; but having in part conquered this disadvantage by the assistance of Mr Braidwood, he made surprising proficiency, becoming a very tolerable classic, and an excellent mathematician.... He fell a victim to his favourite study in 1785 [*recte*: 1786], in consequence of a cold from exposure to night-air in astronomical observatories." (William Turner, *The Warrington Academy* (Warrington, 1957; reprinted from the *Monthly repository*, 1813–15), 75.) On John Goodricke, see Carolyn Gilman, "John Goodricke and his variable stars", *Sky and telescope*, lvi (1978), 400–3, which supersedes the articles in *Dictionary of national biography* and *Dictionary of scientific biography*; Sidney Melmore, "The site of John Goodricke's observatory", *Monthly notices of the Royal Astronomical Society*, lxix (1949), 95–99; and [anon.] "John Goodricke", *ibid.*, xxxv (1912), 435–6 [on the Goodricke portrait now in the possession of the RAS].
18. J. Goodricke, *Journal of astronomical observations begun in Novr 1781* (John Goodricke MSS, North Yorkshire County Council (hereafter: JG MSS)), 1. The first thirty pages of this *Journal*, covering the period to 23 January 1783, have been transcribed from "my old Journal", which does not survive.
19. Pigott to Herschel, 16 February 1783; Herschel MSS, W.1/13.P.29.
20. Goodricke, *Journal*, 47.
21. Letter from Sir Henry Englefield dated 27 November (Pigott, *Journal*, 124).
22. Goodricke, *Journal*, 6. He quickly bought two more eyepieces from Dolland (*ibid.*, 10) but they were not a success.
23. *Ibid.*, 9.
24. *Ibid.*, 22.
25. A copy of the map in Pigott's hand is among the Goodricke MSS.
26. Goodricke, *Journal*, 7.
27. *Ibid.*, 16–18.
28. *Ibid.*, 19–20.
29. Pigott, *Journal*, 171.
30. *Ibid.*, 176.
31. Goodricke, *Journal*, 24–25. The similarity of the phraseology indicates that Goodricke made this entry with Pigott's note before him.
32. *Ibid.*, 28.
33. "Mr E. Pigott has made dilijent researches from what Montanari & Maraldi has observed concer^g Algol ... but it was only a general & vague information" (Goodricke, *Journal*, 42). "P.S. I should have sent you this sooner if I had not been waiting till the last week for the above account of what Montanari & Maraldi had observed concerning this star" (appended by Goodricke to the paper sent to Dr A. Shepherd for publication in *Philosophical transactions*; copy in his hand, JG MSS).
34. Original in Herschel MSS, W.1/13.P.29.
35. "... probably Dr Maskelyne in consequence of my desire, has acquainted you with the phenomenon attending Algol" (Pigott to Herschel, 2 May 1783; Herschel MSS, W.1/13.P.30).
36. W. Herschel, "Observations upon Algol" (manuscript copies in the library of the Royal Society and in Herschel MSS, W.4/9; printed in *The scientific papers of Sir William Herschel*, ed. by J. L. E. Dreyer (London, 1912), i, pp. cvii–cviii).
37. Letters from Banks to Herschel; Herschel MSS, W.1/13.B.6–8.
38. Copy in Herschel's Letter-book, Herschel MSS, W.1/1, 78–79.

39. See his MS copy, JG MSS.
40. Pigott and Goodricke observed together on 13 April (Goodricke, *Journal*, 39), and Pigott returned to York on 7 June (Pigott, *Transit book*, 46); he had written to Herschel from London on 2 May (Herschel MSS, W.1/13.P.30). No doubt Pigott had searched out the Montanari and Maraldi references in London libraries.
41. John Goodricke, "A series of observations on, and a discovery of, the period of the variation of the light of the bright star in the Head of Medusa, called Algol", *Philosophical transactions*, lxxiii (1783), 474–82, p. 482.
42. *Dictionary of scientific biography*, v (New York, 1972), 468.
43. See ref. 36.
44. Copy in Herschel's Letter-book, Herschel MSS, W.1/1, 80–82. On 28 July 1783, Caroline Herschel noted in her *Book of observations* (Herschel MSS, C.1/1.1): "I saw Algol eclipsed...." It may be noted that William Herschel as early as 19 October 1779 had considered "a planet of a very extraordinary size in the system of worlds, round which a Sun ... may move in about a years time" in order to explain variations in brightness in Mira Ceti (as in its supposed orbit it approached to and receded from the observer) (*ibid.*, W.4/1.1, f.32).
45. John Goodricke, "On the period of the changes of light in the star Algol", *Philosophical transactions*, lxxiv (1784), 287–92.
46. Draft in JG MSS.
47. Goodricke, *Journal*, 49.
48. *Ibid.*, 93.
49. In the pleasant phrase of Nathaniel Pigott, letter to Maskelyne, May 1784 (N. Pigott MSS, North Yorkshire County Council, "Mathematical correspondence 1783, 1784, 1786").
50. Goodricke, *Journal*, 40.
51. Pigott, *Journal*, 311.
52. A convenient source for knowledge of variable stars at the beginning of the present century is J. E. Gore, *The stellar heavens* (London, 1903), chap. 3.
53. In his major survey, "Observations and remarks on those stars which the astronomers of the last century suspected to be changeable", *Philosophical transactions*, lxvi (1786), 189–219; see p. 196.
54. As shown by scraps of dated observations, EP MSS.
55. Quoted from a scrap of a draft letter, EP MSS.
56. "Mr Goodricke informs us, that he is very happy in London, and mentions the polite reception at Greenwich and the civilities he received from you with great pleasure"(letter of N. Pigott to Maskelyne, 2 April 1784; N. Pigott MSS, "Mathematical correspondence 1783, 1784, 1786"). Goodricke makes public acknowledgement to Maskelyne at the beginning of his paper on δ Cephei, "A series of observations on, and a discovery of, the period of the variation of the light of the star marked δ by Bayer, near the Head of Cepheus", *Philosophical transactions*, lxxvi (1786), 48–61. Maskelyne, in writing to N. Pigott on 22 April 1784, remarks: "I am glad that Mr Goodricke applies himself so much to his studies, which may in time enable him to distinguish himself in the mathematical as well as philosophical world" (original in Royal Astronomical Society archives, Pigott 58).
57. Goodricke, *Journal*, 79.
58. John Goodricke, "Observations of a new variable star", *Philosophical transactions*, lxxv (1785), 153–64.
59. *Ibid.*, 163; the draft in the Goodricke MSS shows the last phrase to be an afterthought.
60. Draft, JG MSS.
61. Edward Pigott, "Observations on a new variable star", *Philosophical transactions*, lxxv (1785), 127–36, p. 127.
62. *Ibid.*, 134.
63. *Ibid.*, 134–5.
64. Goodricke, *Journal*, 94.
65. *Ibid.*
66. Goodricke, *op. cit.* (ref. 56), 60–61.
67. Goodricke to Herschel, 17 January 1785; Herschel MSS, W.1/13.G.14.
68. As shown by the collected observations at the back of his *Journal*.
69. See ref. 17.
70. *Op. cit.* (ref. 53).
71. *Ibid.*, 215.

72. Cited above, ref. 8.
73. On which they separately reported in *Philosophical transactions*, lxxvi (1786), 384–8 and 389.
74. E. Pigott, *op. cit.* (ref. 8), 424.
75. Pigott, *Journal*, 337.
76. Their movements are shown by the place entries in Edward's *Journal*.
77. The loss "took away the pleasure I ever had in astronomical pursuits, until revived again by accidentally seeing a Comet" on 14 January (*Transit book*, 77). Shortly before his mother's death, Edward had come across a copy of Thomas Wright's *An original theory or new hypothesis of the universe* (London, 1750), and it is interesting that he studied it closely (unlike William Herschel) and concluded: "the whole of this book is replete with new matter & luminous thoughts" (*Journal*, 356–7).
78. The instrument he now used is described in detail in the *Transit book*, 79, 88–89, 97–98, 115–16.
79. Pigott, *Journal*, 199.
80. *Ibid.*, 257.
81. *Ibid.*, 384.
82. *Ibid.*, 388.
83. Edward Pigott, "On the periodical changes of brightness of two fixed stars", *Philosophical transactions*, lxxxvii (1797), 133–41.
84. Pigott, *Journal*, 321.
85. *Ibid.*, 336 bis.
86. *Ibid.*, 343.
87. *Ibid.*, 392.
88. *Ibid.*, 393.
89. W. Herschel, "On the nature and construction of the Sun and fixed stars", *Philosophical transactions*, lxxxv (1795), 46–72, p. 68.
90. Pigott, *Journal*, 389.
91. *Philosophical transactions*, lxxxvi (1796), 452–82, p. 458. The suggestion of stars being flattened by quick rotatory motions goes back to P. L. M. de Maupertuis, *Discours sur les différentes figures des astres* (Paris, 1732).
92. Flamsteed's *Atlas* (French edn, Paris, 1776) in EP MSS, facing p. viii.
93. Loose sheets inserted in Pigott's *Journal*.
94. Edward Pigott, "An investigation of all the changes of the variable star in Sobieski's Shield, from five year's observations, exhibiting its proportional illuminated parts, and its irregularities of rotation; with conjectures respecting unenlightened heavenly bodies", *Philosophical transactions*, xcv (1805), 131–54.
95. *Ibid.*, 144.
96. *Ibid.*, 145.
97. *Ibid.*, 152.
98. *Ibid.*, 152–3.
99. *Ibid.*, 154.
100. Cited, from a manuscript in his possession, by Gavin de Beer, *The sciences were never at war* (London, 1960), 138.
101. *Ibid.*, 161.
102. Melmore, *op. cit.* (ref. 4), 285.
103. De Beer, *op. cit.* (ref. 100), 154–5.
104. Letter to Herschel, 30 October 1806; Herschel MSS, W.1/13.P.38.
105. He wrote to Herschel on 2 February and 28 September 1807 and 31 August 1811 (*ibid.*, P.39–41) to draw his attention to comets, and careful drawings of the 1811 comet survive in the Pigott MSS.
106. Letter to Herschel, 24 August 1817; Herschel MSS, W.1/13.P.42.
107. Cited by Kopal in his article on the Pigotts in *Dictionary of scientific biography*.
108. Pigott to John Herschel, 10 May 1821; Herschel Papers, Royal Society. I owe this reference to Dr S. S. Schweber.
109. See ref. 93.

5. Herschel's Determination of the Solar Apex

Throughout the eighteenth century astronomers and cosmological speculators considered the possibility that the Sun and the solar system are moving through space;[1] indeed, on Newtonian gravitational theory it was not to be expected that the Sun or any star would be at rest despite the gravitational pulls of the other stars. How the motion of the Sun would reveal itself to us by generating a pattern of proper (or individual) motions in the nearby stars was explained by Tobias Mayer in a lecture in 1760:

> For if the Sun and all the planets together with it, including our own home the Earth, were moving straight towards some region, all the stars which appear in that region would seem to be gradually separating from each other one by one, and those which are in the opposite part of the sky would seem to be joining up; just as when you are walking through a wood the trees which are in front of you seem to be separated and those which are behind you seem to be joined together.[2]

In the same lecture, Mayer listed[3] the modern positions of some eighty stars together with their positions as recorded half-a-century earlier:

> ... there are no fewer than fifteen or twenty which, in the now sufficiently perceptible space of fifty years, were found to be endowed with their own proper motion.

However, Mayer had a healthy scepticism concerning the alleged changes in position of most of his stars:

> In order that it may more easily be judged to what extent both [pairs of coordinates] agree or disagree, I have noted the difference in separate columns. This indeed, as is clear, is so slight in [the case of] most of the stars that, within the limits of observational error, it appears that they are really fixed and subject to no movement which can be perceived by our sense organs. However, in some cases the differences are a little greater than can be attributed to the imperfection of the instruments. Consequently, it would appear to be not improbable that those which exceed 10 or 15 seconds of arc, and which on this account are distinguished from the rest, indicate a certain degree of movement. But in the case of Arcturus, Sirius, Procyon, Pollux, [α Aquilae, γ Piscium] and certain other stars, the difference between their former and present-day positions is so great, that there can be no doubt whatever concerning their movement.

Mayer, in fact, had for the first time provided data sufficiently numerous to reveal a pattern such as he had described; but, declared Mayer, no such pattern was present: "... these motions of stars are not bound by a common law of this kind. ... Perhaps the true and genuine reason for these movements will still remain unknown for many centuries."[4]

Such pessimism was echoed by Lalande in a memoir[5] published in 1779: the solar motion would be detected only after a lapse of centuries, when the Sun

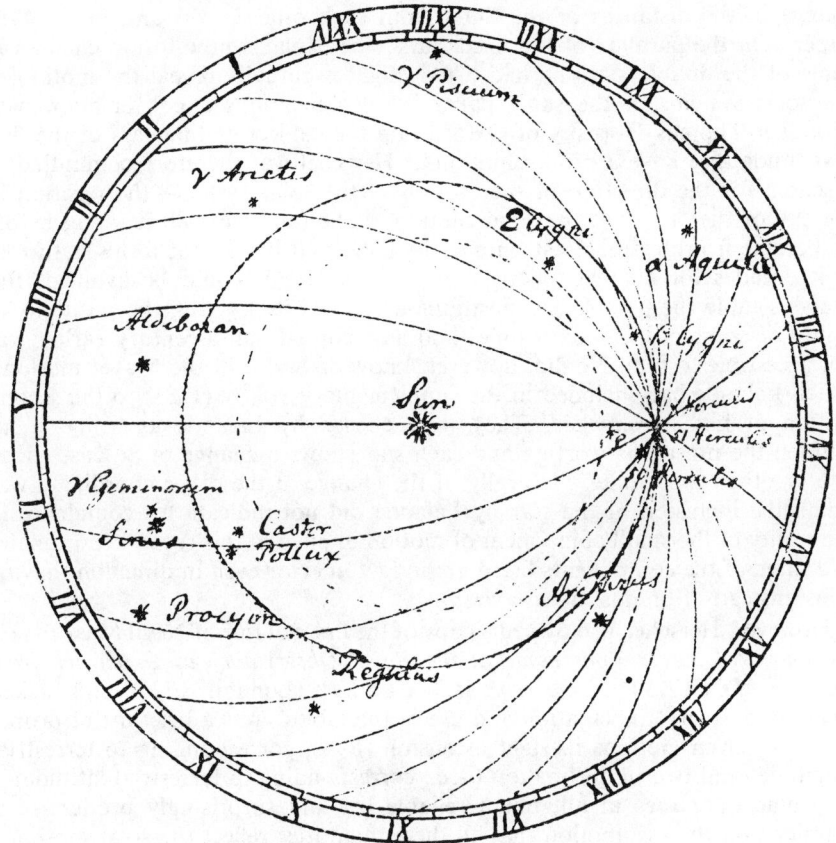

FIG. 1. Sketch by Herschel for the first draft of his paper "On the proper motion of the Sun and solar system", showing the projection onto the equatorial plane of the stars whose proper motions are listed by Maskelyne or Lalande (Sirius is in the southern hemisphere). The curves radiating from λ Herculis are great circles along which stars will appear to move away from λ Herculis if that star is the direction towards which the solar system is moving. At this stage Herschel was also considering 50, 51 and ρ Herculis as alternative directions for the solar motion, for reasons discussed below. Note that the sky is reversed from that seen from Earth, suggesting that Herschel may have been working with a celestial globe. (Royal Astronomical Society W. Herschel MSS 4.19.2, reproduced courtesy of the Royal Astronomical Society.)

arrived perceptibly closer to one region of the sky than to the opposite region. Yet within four years William Herschel had announced[6] a direction in which the solar system was moving that is close to the best modern values. His achievement owed nothing to his pre-eminence as an observer; it was a desk-exercise, using published data available to any astronomer. How did Herschel, whose grasp of mathematics was minimal, discern a pattern where others had failed?

Herschel had cut his astronomical teeth on James Ferguson's *Astronomy*, and there he would have read that "If our Solar System changes its place with regard to absolute space, this must in process of time occasion an apparent change in the distances of the Stars from each other...";[7] and in his 1781 paper "On the parallax of the fixed stars"[8] he explains how future changes in some of the double stars he had listed might eventually reveal the motion of the solar system. In the same paper he also mentions a conversation with Professor Thomas Hornsby of Oxford "on the subject of the stars of the first magnitude that have a proper motion".[9] Herschel was therefore committed to a search for the direction of the motion of the solar system—the position in the sky of the 'apex' of the solar motion, as he (and we) call it—long before he became a professional astronomer in 1782, and it was not in his nature to wait decades for changes in his double stars, if this could be avoided. But Mayer's table (first published posthumously in 1775[10]) giving the positions of eighty or so stars in modern times and as recorded half-a-century earlier was not accessible to him. He did, however, know of twelve of the Mayer motions, which Lalande had included in the supplementary vol. iv (1781) to the second edition of his *Astronomie*.[11] They were chosen by Lalande as being in his opinion the most trustworthy, and each star shows a change of at least 18" in one or other coordinate. Naturally, if the change in the other coordinate was small, the inclusion of the star by Lalande did not indicate his confidence in the reality of the small component of motion implied in this second coordinate; and three of the components listed are in fact in error even in direction (having 'plus' instead of 'minus' or *vice versa*).

However, Herschel also owned a copy of the first volume of Nevil Maskelyne's *Astronomical observations made at the Royal Observatory at Greenwich from the year MDCCLXV to the year MDCCLXXIV* (London, 1776), and tucked away on p. iv of "Explanation and use of the tables" was a brief list of proper motions: seven motions in right ascension (R.A., corresponding to terrestrial longitude) and two in declination (dec., corresponding to terrestrial latitude)— only nine quantities in all, but a reliable list and surprisingly productive if examined on the assumption that all these quantities reflect the solar motion.

No doubt Herschel worked with both lists simultaneously,[12] but he encourages us to think that he began with the Maskelyne motions and only afterwards moved on to those of Lalande, and it is helpful to acquiesce in this. Herschel had no difficulty in appreciating that, if the solar system is moving towards a given apex, and if the opposite point in the sky is the 'antapex', then as a result of the solar motion every star will appear to move (perceptibly or otherwise) along the great circle passing through the apex, the star, and the antapex, in the direction away from the apex. (Figure 1 is a sketch by Herschel with these circles drawn for the apex he eventually derives; note that the sky is drawn as viewed from the 'outside', suggesting that Herschel, sensibly, worked with a celestial globe.) The data of proper motions, however, were given in components in R.A. and dec. Fortunately, for a given apex the directions of expected changes in R.A. are easy to compute: if we divide the sky into two halves by the great circle passing through the poles and the apex, then all the stars in one half will increase in R.A. while those in the other half will decrease. Changes in declination are more subtle, and the mathematics was beyond the powers of Herschel (and of most modern astronomers!); but more of this anon.

Figure 2 shows in schematized form the directions of motion in R.A. of Maskelyne's seven stars. And we see that all the directions will be as expected

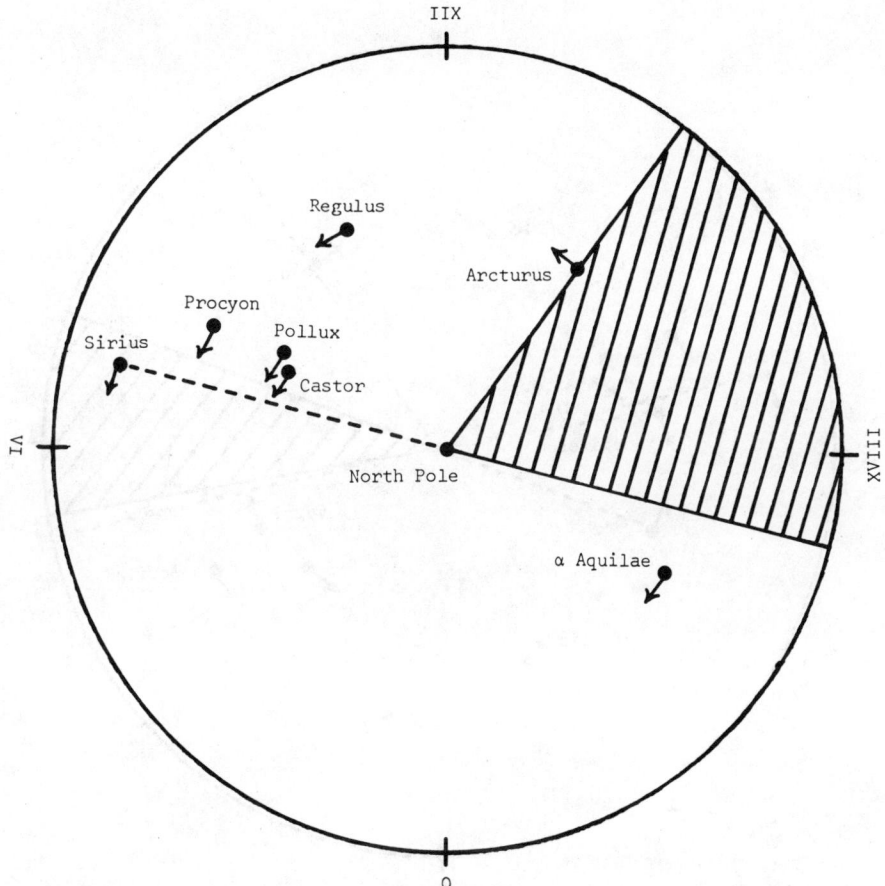

FIG. 2. Schematic figure showing the directions (but *not* the quantities) of the components in R.A. of proper motions listed by Maskelyne. If the solar apex is in the shaded region, then the motion of the solar system will produce in all seven stars apparent motions whose components in R.A. have the directions shown.

if the solar apex has R.A. between the R.A. of Arcturus and the R.A. of a star opposite to Sirius.

Figure 3 shows the same data for the Lalande stars (namely, six of the Maskelyne stars, fortunately all with directions the same as those given by Maskelyne, plus six further stars). It is important to note that Herschel simply accepts the data as true, even though the change in fifty years in the R.A. of γ Geminorum was given as only 8″, and of β Cygni and of Aldebaran as a mere 3″ each. However, if we regard ourselves as limited to the region already derived from the Maskelyne stars, then we see that we must accept γ Arietis and β Cygni as exceptions; but the change in Aldebaran enables us to restrict much more severely the region within which a solar apex will account for the direction of change in the remaining stars, as does the change in γ Geminorum

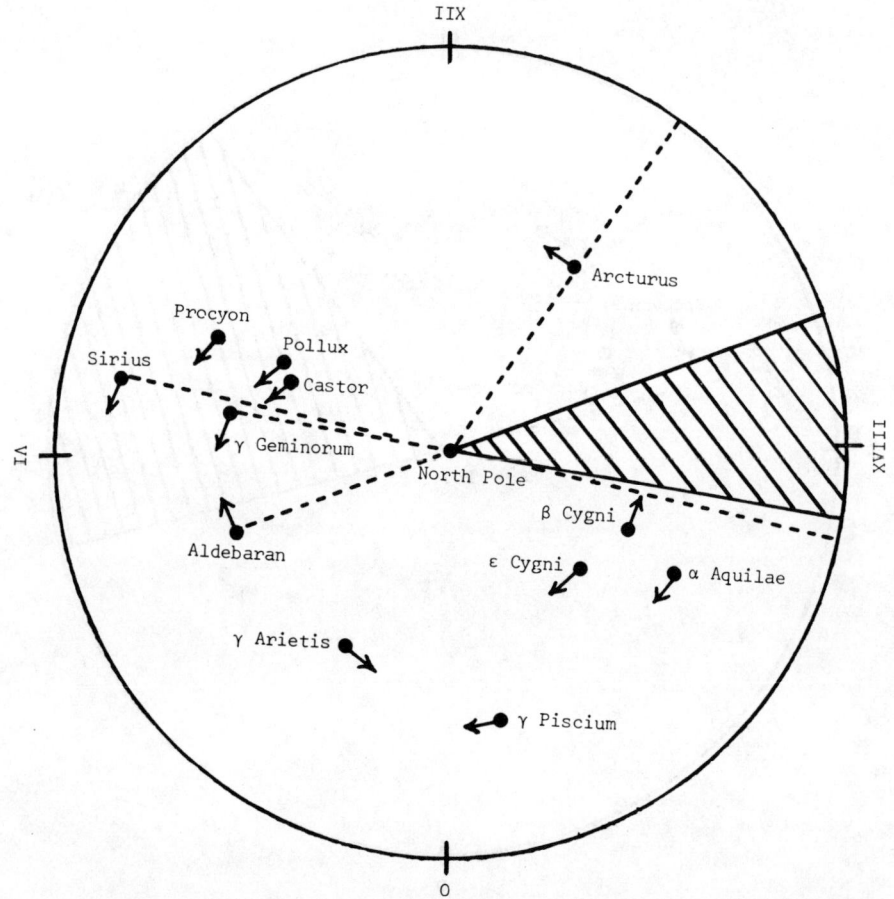

FIG. 3. Schematic figure showing the directions (but *not* the quantities) of the components in R.A. of the changes listed by Lalande. If the solar apex is in the shaded region, then the motion of the solar system will produce in all the stars except γ Arietis and β Cygni, apparent motions whose components in R.A. have the directions shown. But note that whereas Lalande had selected these stars from Mayer's list because in each case, when the changes in R.A. and dec. were combined, the resulting shift in position over the fifty years had been substantial and convincing, we find Herschel giving as much weight to the highly questionable 'motions' of 3″ in R.A. for Aldebaran and β Cygni in the half-century as he does to the change of 1′11″ recorded in the R.A. of Arcturus. (Mayer considered that changes below 10″ or 15″ might well be the result of observational errors.) Without Aldebaran, the shaded region would be much larger.

to a lesser extent. The (restricted) region now has bounds in R.A. opposite to γ Geminorum and Aldebaran—but the improvement is entirely due to our (that is, Herschel's) accepting at face value 'motions' far below the level of observational error. Had the 3″ change recorded for Aldebaran happened to have the opposite sign, this would have been taken by Herschel to imply motion in the opposite direction, and we would now be left with the other half of the region derived from the Maskelyne data!

Herschel's Determination of the Solar Apex

We now turn our attention to the much more difficult question of how to infer the declination of the solar apex from the motions in declination of the Maskelyne and Lalande stars. Maskelyne gave only two such motions, both large and well-attested: Arcturus (in the northern hemisphere) was moving southwards, as was Sirius (in the southern hemisphere, but less than 17° below the equator). Herschel perceived that within the region shaded in Figure 3, an apex higher in the sky than Arcturus would be necessary. Figure 4 indicates the directions of motion in declination of the Lalande stars in the northern hemisphere (that is, other than Sirius), though the change for γ Piscium is listed as only 7″, that of α Aquilae as only 4″, and for Castor Herschel has accepted an absurd 1″. We can see that with an apex very roughly between Arcturus and the North Pole, most of the stars are satisfactorily moving away from the apex. γ Piscium is a clear exception; but what are the implications of the northward motions of β and ϵ Cygni? Herschel thought that an apex within the shaded region of Figure 3, and with declination north of Arcturus and south of β and ϵ Cygni, would produce the required southward motion of Arcturus *and* the northward motions of β and ϵ Cygni: that is, an apex within the shaded region in Figure 4. Nearly central to this region is the star λ Herculis, which Herschel therefore provisionally fixed upon as the approximate position of the solar apex. He accordingly submitted to the Royal Society his paper "On the proper motion of the Sun and solar system", in which he claimed that an apex near λ Herculis would satisfy all the Maskelyne and Lalande data with only three exceptions: γ Arietis and β Cygni in R.A., and γ Piscium in declination. "With regard to the change of declination," he wrote, "we see that β and ϵ Cygni should go towards the north pole as lying between the Pole and the point towards which the sun is supposed to move; and that all the rest ought to go towards the south, or encrease their n[orth] polar distance."[13]

To satisfy the data with only three exceptions was a remarkable achievement. But with the apex at λ Herculis, the southward motions in declination of Sirius and Arcturus would be small, whereas Maskelyne and Lalande were agreed that the motions were large, and indeed that of Arcturus had been notoriously so for decades. Herschel felt these well-established southward motions of two very bright stars must be given great weight, and therefore he considered modifying the position of the solar apex, even though this would turn the motions in dec. of β and ϵ Cygni into additional exceptions:

> If the change in declination of β & ϵ Cygni be thrown on their own proper motion, rather than that of the Solar-System, we shall be able to make other quantities agree much better with observation, by supposing the Sun to move towards a more northern star, such as ρ Herculis. It seems the motion of Arcturus and Sirius is too great to permit the Apex to lie so low as either λ or the 51st Herculis. If it were raised to the parallel of the 50th of the same constellation, we should only throw out β Cygni, which is a star from other reasons much to be suspected of a real proper motion; but even this elevation would hardly suffice: I would rather consider Arcturus and Sirius as the leading marks to regulate that point. It can not be said that I act arbitrarily in assigning to ϵ and β Cygni motions at pleasure, as best suits my hypothesis; for since proper motions have actually been observed, if they are not to be accounted for from

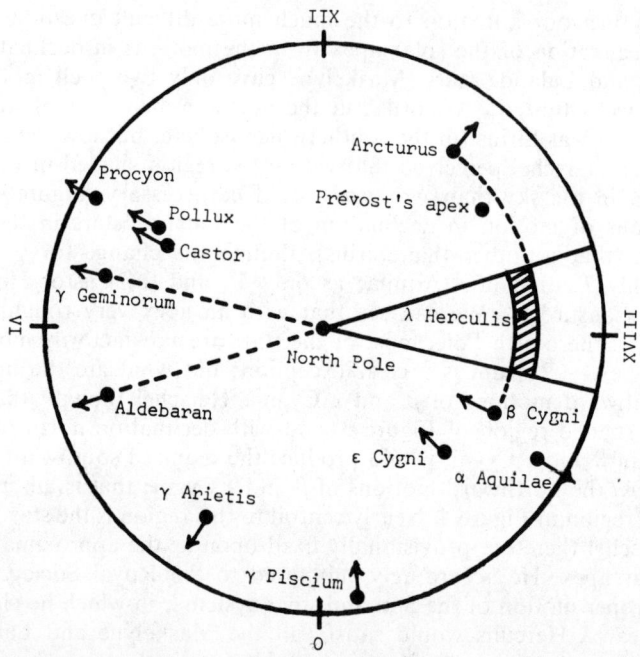

FIG. 4. Projection of the northern hemisphere onto the equatorial plane, showing the directions (but *not* the quantities) of the components in dec. of all the changes listed by Lalande (except for Sirius, which is in the southern hemisphere). Herschel at first thought that, if the solar apex is to be selected within the shaded region in Fig. 3, then an apex more northerly than Arcturus and more southerly than β Cygni (that is, in the region here shaded) will produce in all the stars except γ Piscium, apparent motions whose components in dec. have the directions shown. Note that λ Herculis is centrally placed in the region so defined. Maskelyne pointed out that β and ε Cygni were in fact additional exceptions since they should move southerly if the apex is at λ Herculis.

Also shown is the solar apex proposed by Prévost (R.A. 230°, dec. 25°N). Had the negligible change of 3″ in the R.A. of Aldebaran been listed with the opposite sign, inspection of Fig. 3 suggests that Herschel's apex might well have nearly coincided with Prévost's.

that of the Sun, they must needs be in the stars themselves; and as in this hypothesis I would intirely be directed by facts, it is highly proper, where contrary motions appear, to lean to the side which has the strongest arguments in its favour.[14]

Figure 1, reproduced here for the first time, is a sketch drawn for this first draft of Herschel's paper, and shows the four stars in Hercules mentioned in this paragraph.

Herschel, then, far from being rigidly committed to λ Herculis as the solar apex, was now inclined to place the apex as far north as ρ Herculis, even though this meant turning the motions in declination of β and ε Cygni into exceptions. In fact Herschel had made a mathematical blunder: they were exceptions even with the apex at λ Herculis, as Maskelyne told him a few days after his paper was read to the Royal Society:

> In addition to the remarks I sent you on your paper on the motion of the solar system, let me add, that you seem to have made a mistake in calcu-

TABLE 1. Stars listed by Mayer as having, according to the observations, altered position in R.A. by at least 10″ in fifty (or in certain cases forty-four) years.

Star	R.A. °	Change in R.A. ″	Notes	Star	R.A. °	Change in R.A. ″	Notes
β Cet	8	+32		ζ Hya	131	−23	
γ Ari	25	−14	1*	α Leo	149	−16	
α Ari	28	+10		ε UMa	191	−33	
δ Cet	37	+15		α Boo	211	−71	
γ Cet	38	−14	1*	[Prévost's apex]			
α Cet	42	+16		β Her	245	+14	3
β Per	43	−10	1*	[Herschel's apex]			
α Per	47	+16		γ Dra	268	+12	
[Prévost's antapex]				α Aql	295	+32	
γ Eri	57	+14	2*	γ Cyg	303	−13	1*
α Aur	75	+11	2*	ε Cyg	309	+18	
[Herschel's antapex]				γ Cap	322	+19	
β Tau	78	−11		ε Peg	323	−14	1*
μ Gem	92	−16		δ Cap	323	+24	
α CMa	99	−37		α Aqr	328	+13	
α Gem	110	−24		ζ Peg	336	−20	1
α CMi	112	−33		α PsA	341	+21	
β Gem	113	−48		β Peg	434	+12	
ρ Pup	119	−13		γ Psc	346	+53	
ι UMa	131	−54		β Cas	359	+34	

Notes: 1. Exceptions for Herschel and Prévost.
2. Exceptions for Prévost but not Herschel.
3. Exceptions for Herschel but not Prévost.
*These stars have changes listed as below 15″ and are therefore disregarded by Prévost (see ref. 18), who is left with one single exception and twenty-one successes.
In certain stars the changes are over forty-four years rather than fifty: β Cet, δ Cet, γ Cet, γ Eri, ρ Pup, ι UMa, ζ Hya, ε UMa, β Her, γ Cyg, ε Cyg, ε Peg, ζ Peg, β Cas.

lating the effect of the supposed motion towards λ Herculis on the declination of the stars; for by that motion β & ε Cygni will be carried a little southward, instead of northward as you suppose; and therefore contrary to the observations you adduce from M. DeLalande. I find too that upon your hypothesis the motions of Arcturus & Sirius southward will be very small, contrary to observation; nor can the matter be mended by altering the apex of the solar motion as you call it; as that would make the motions of the other stars to differ still more from the hypothesis than they do at present.[15]

Maskelyne was no hostile critic: "Your hypothesis struck me when I wrote the preface to my Tables & is probable; yet I could point out to you a star of the 3ᵈ magnitude which has a considerable proper motion in contradiction to it."[16]

The simplicity on which Herschel's progress had depended was rapidly being obscured. His friend, the amateur observer Alexander Aubert, contributed most to this when he sent Herschel a copy of Mayer's *Opera inedita* on the very day his paper was read to the Royal Society.[17] Herschel was now confronted with the eighty real or imaginary changes of position listed by Mayer. It would be impossible to cope with so much information by the simple graphical methods Herschel had so far used; and Mayer himself had ascribed the majority of the recorded changes to instrumental error. In general, Herschel inclined to distinguish sharply between facts and their interpretation—in his work on nebulae a reliable observation was of enduring value, to be produced and paraded for inspection whenever it served a purpose. He was ill at ease when the data themselves were suspect. Also, once committed to a position (as in the paper as submitted and read to the Royal Society), he was reluctant to reconsider. Therefore, instead of beginning his investigation over again, he made a half-hearted attempt to check the existing results against the new data, even though he would now reject as 'insensible' changes of less than 10″ where earlier such changes, especially in the R.A. of Aldebaran, had been taken as evidence of decisive importance. And so he composed a Postcript in which he claimed to check the additional Mayer changes of 10″ or more in either coordinate against an apex in the region of λ or ρ Herculis, and produced a list of stars whose 'motions' were compatible with such an apex and a list of those whose 'motions' were not. There are signs of special pleading; but in general Herschel was right to claim that, more often than not, the new data favoured an apex near λ or ρ Herculis.

In R.A. we see this from Table 1, which lists all the Mayer changes in R.A. of at least 10″. Herschel claims both α Aur and β Tau as favouring his hypothesis, which specifies his apex as about 257° in R.A. Granted this, there is a striking run of 'successes' in the twelve stars from β Tau to α Boo, all of which had decreased in R.A.: there *is* therefore a reasonably obvious pattern in the Mayer data in R.A., but Mayer simply failed to notice it. In fact, Herschel can claim that of the 36 changes listed in the table, only seven are exceptions, while the remaining 29 count as 'successes'. The success rate is undeniably impressive.

However, Herschel's is not the only apex with a fine success rate. Later in 1783, Pierre Prévost also published an estimate of the solar apex derived from

TABLE 2. The number of 'exceptions' when the solar apex is chosen to have R.A. near the values proposed by Herschel and Prévost.

Star	R.A. of star °	R.A. of apex $A°$	Resulting no. of 'exceptions'
		$A < 227$	9
α Per	47 [+180 = 227]		
		$227 < A < 237$	8
γ Eri	57 [+180 = 237]		
		$237 < A < 245$	7
β Her	245		
		$245 < A < 255$	8
α Aur	75 [+180 = 255]		
		$255 < A < 258$	7
β Tau	78 [+180 = 258]		
		$258 < A < 268$	8
γ Dra	268		
		$268 < A$	9

Note: These exceptions are counted on the assumption that changes in R.A. of 10″ in fifty (or forty-four) years are to be considered significant. If however we ignore changes of less than 15″, then there are no data of significance between 227° [α Per] and 272° [μ Gem]. If we ask for changes of at least 30″, then we have the range from 211° [α Boo = Arcturus] and 279° [α CMa = Sirius], as derived in Fig. 2 from the Maskelyne data!

the Mayer data,[18] and his apex (see Figure 4) was 230^0 in R.A. As we see from Table 1, on the same test he could claim a success rate of 28 out of 36—or 21 out of 22, since he disregarded changes below 15″. In fact the Mayer data in R.A. does not permit a clear choice of position for the solar apex, which could reasonably be anywhere in the range 227^0 (or less) and 268^0 (or more; see Table 2). If, as I have argued, Herschel originally selected λ Herculis (R.A.: 260^0) because it was midway between the bounds defined by Aldebaran ($65^0 \rightarrow 245^0$) and γ Geminorum ($96^0 \rightarrow 276^0$), then we realize that he was permanently diverted away from the equally satisfactory values in the range from (say) 227^0 to 245^0 because he gave great weight to the absurdly small difference of 3″ in Aldebaran's position in R.A. as recorded by Römer in 1706 and Mayer in 1756. *It is because of these three seconds of arc that Herschel derived an apex close to the best modern values*, while Prévost's is nearly thirty degrees away in R.A.

In declination the situation is less clear-cut, though Herschel is right to claim that in general the verdict is favourable. However, one fact of importance was among the Mayer data: a northward change in declination of the bright star α Lyrae of 14″ in fifty years, a quantity just to small for inclusion by Lalande. Since α Lyrae is quite close to ρ Herculis and differs from it by only slightly more than 1^0 in declination, its expected motion would be almost entirely in R.A. if the apex were at ρ Herculis; the northward change listed by Mayer therefore favoured a more southerly apex, such as λ Herculis, some 11^0 south of ρ Herculis. It is almost certainly for this reason that Herschel altered his original draft to eliminate all reference to ρ Herculis, and it is to α Lyrae that he refers when he added: "From the additional testimony of other capital stars considered in the postscript it now appears, that the point λ Herculis is probably as well chosen as any we can fix upon in that part of the heavens."[19] But he was ill-at-ease with the flood of new data, and here for the present his investigation ended: "I forbear entering too much into refined consideration; what we are chiefly to determine at present is, an outline or sketch of what many repeated, and farther extended, observations must ripen so far as in time to enable us to apply more particular calculations."[20]

Conclusions

(1) The changes in R.A. listed by Mayer show a clear pattern, favouring a solar apex with R.A. within the approximate range 227^0 (or less) to 268^0 (or more).

(2) Herschel worked first with the limited data provided by Maskelyne and Lalande. Because he accepted at face-value the change of 3″ in fifty years in the R.A. of Aldebaran, he was led to propose an apex in the upper part of the range. This apex was compatible with the majority of the full Mayer data when they became available.

(3) In declination Herschel's choice of solar apex was influenced by the large southward motion of Arcturus, and later by the northward motion of α Lyrae. An apex thus defined is compatible with the majority of the full Mayer data.

(4) A solar apex in the region of Hercules and Corona Borealis could reasonably be inferred from the Mayer data; but the extraordinary closeness of Herschel's

apex to the best modern values owed much to his illicit use of the listed change of 3″ in the R.A. of Aldebaran.

REFERENCES

1. Going back at least to Robert Hooke (*Posthumous works*, ed. by Richard Waller (London, 1705), 506), and including Wright, Kant and Lambert.
2. *Tobias Mayer's Opera inedita*, trans. and ed. by E. G. Forbes (London, 1971), 112. Forbes suggests (p. 44) that Herschel succeeded where Mayer failed because Herschel opted to consider only Mayer's brightest stars, believing them to be the nearest; but we shall see that this is not the explanation.
3. The list is omitted in the English translation. See ref. 10 below.
4. *Op. cit.*, 110–12; translation by Prof. Forbes.
5. J. de La Lande, "Mémoire sur les taches du Soleil, et sur sa rotation", *Mémoires de l'Académie royale des Sciences* for 1776 (Paris, 1779), 457–514, pp. 513–14.
6. W. Herschel, "On the proper motion of the Sun and solar system; with an account of several changes that have happened among the fixed stars since the time of Mr. Flamsteed", *Philosophical transactions*, lxxiii (1783), 247–83.
7. James Ferguson, *Astronomy explained upon Sir Isaac Newton's principles* (2nd edn, London, 1757), 237.
8. W. Herschel, "On the parallax of the fixed stars", *Philosophical transactions*, lxxii (1782), 82–111 (read 6 December 1781).
9. *Ibid.*, 98.
10. *Opera inedita Tobiae Mayeri*, I, ed. by G. C. Lichtenberg (Göttingen, 1775), 80–81.
11. Three volumes of the second edition appeared in Paris in 1771, and the fourth (supplementary) volume in 1781. By that time Mayer's *Opera inedita* had appeared and Lalande was able to include a select list of Mayer 'proper motions' in his §2756.
12. Most of his rough sheets were destroyed after his death, but a page of jottings taken from Lalande survives, after which Herschel makes the (careless) comment that "All these motions [listed by Lalande] seem to favour a direction of ☉ towards α Draconis" (Royal Astronomical Society W. Herschel MSS, 7/14.27), which suggests he had not yet analysed the Maskelyne data.
13. Royal Astronomical Society W. Herschel MSS, 4/19.1.
14. *Ibid.*
15. Maskelyne to Herschel, 15 March 1783 (Royal Astronomical Society W. Herschel MSS, 1/M.22). On the back of the letter Herschel has drawn a sketch to try and verify Maskelyne's assertion as to the effect on β Cygni of a solar apex at λ Herculis.
16. *Ibid.*
17. Aubert to Herschel, 7 March 1783 (Royal Astronomical Society W. Herschel MSS, 1/A.11): "I sent you yesterday by the Windsor Coach from the Bell Savage Inn, Ludgate Hill, a Box, directed to you at Datchet, containing my *Atlas Coelestis*, the book Mr Cavendish sent me for you & Mayer's *Opera Inedita*.... We finished last night your Paper on ye motion of our whole System it pleased me much & I think you bring very strong proofs in favour of your Ideas the paper seemed to give general satisfaction it was much commended after the meeting & I wish you joy of the rapid increase & accumulation of your fame...."
18. First published as "Ueber die Fortrückung unsers Sonnen-Systems", *Astronomisches Jahrbuch für das Jahr 1786* (Berlin, 1783), 259–60. Prévost's apex had the same declination as Herschel's, namely 25°.
19. Herschel, "On the proper motion of the Sun and solar system", footnote added on p. 273.
20. *Ibid.*, 279 (in the Postscript).

B. A CENTURY OF SPECULATIVE COSMOLOGIES

1. *Introduction*

Until William Herschel built his giant telescopes in the 1780s and used them to reach out for great distances into space, the large-scale structure of the universe – cosmology – was a field for the occasional philosophical or theological speculator rather than for the scientific astronomer. Newton himself seems to have wasted little time on cosmology until 1692 when he was directly asked by the Rev. Richard Bentley whether the number of stars was finite or infinite, and whether the structure of the universe showed the hand of God at work. He concluded (see Section B, chap. 2) that the study of the stars revealed the same two-fold activity of Providence as did the study of the solar system. The remarkable structure of the solar system with its Sun, planets and comets taught the natural philosopher that Providence had endowed Man's immediate environment with a long-term stability that could not have come about by accident. Yet, eventually, even so stable a system must be threatened with collapse and "want a reformation",[1] and then Providence would directly intervene to restore stability and so demonstrate to Man the *continuing* concern of God for His creation. Similarly, the huge distances separating one star from another, and the regularity with which the stars were distributed in space, guaranteed long-term stability to the system of the stars and demonstrated that the system had been planned by Providence; yet here, too, collapse must eventually threaten, and Providence would similarly intervene to restore stability and demonstrate to Man the continuing concern of God for His creation.

The regularity that Newton's (unpublished) analysis revealed in the system of stars was far from obvious to his younger contemporaries, to whom the apparent *lack* of order among the stars was a cause for concern. Newton's successor at Cambridge, the deeply-religious William Whiston, told his students that "There may be a certain orderly and harmonious Disposition of the Fixed Stars amongst themselves, when they are beheld from some other proper Place, altho' that Order appears not when they are seen from this Earth [just the same as happens with the planets]: Or this Order may consist in certain beautiful Proportions, fitted to the several Systems, which are wholly unknown to us".[2] Similarly, for William Derham, writing in *Astro-theology,* there is a "great Parity and Congruity observable among all the works of the Creation; which have a manifest harmony, and great agreement with one another".[3] Accordingly, "these several Systemes of the Fixt Stars, as they are at a great and sufficient Distance from the Sun and us; so they are imagined to be at as due and regular Distances from one another. By which means it is, that those multitudes of Fixt

Stars appear to us of different Magnitudes, the nearest to us large; those farther and farther less and less".[4] Modern telescopes, he says, have revealed many more stars than were formerly known,

> and all these far more orderly placed throughout the Heavens, and at more and due agreeable distances, and made to serve to much more noble and proper ends .... [The stars] are not set at random, like a Work of Chance, but placed regularly and in due order ... they look to us, who can have no regular prospect of their positions, as if placed without any order: like as we should judge an army of orderly, well disciplined soldiers, at a distance, which would appear to us in a confused manner, until we came near, and had a regular prospect of them, which we should then find to stand well in rank and file.[5]

Whiston's *Astronomical principles of religion, natural and reveal'd* was one of the works closely studied by the young Thomas Wright of Durham (see Section B, chap. 3), an autodidact who attempted to incorporate what he knew of scientific astronomy – that is, of the observed regions of space – into his theological vision of the universe as a whole, according to which the Sun and the other stars orbit around the Divine Centre. We have seen briefly in Section A, chap. 1 how Wright achieved a partial understanding of the Milky Way, and it is with this that his name has been associated by historians; but this was an accidental byproduct of his main purpose, which was to reconcile science and religion in the context of a stable universe in which fire has a special role[6] and changes are cyclic.

It was Immanuel Kant who rescued Wright from threatened oblivion (see Section B, chap. 3), although in the short term Kant's own publication was effectively still-born because of the bankruptcy of his publisher. But Kant's *Allgemeine Naturgeschicht und Theorie des Himmels* (1755) contains much more than the first statement of the Milky Way as a disk-shaped system of stars each of which is in orbit. Kant pictures for us the universe as beginning with an initial chaos. In one region we envisage the density of matter as being unusually high, and there gravitational forces are unusually great and so order, structure, is first formed out of chaos. As time goes on the boundary of order extends further outwards, but meanwhile gravity works on the newly-formed order to reduce it to chaos once more:

> Let us now proceed to trace out the construction of this Universal System of Nature from the mechanical laws of matter striving to form it. In the infinite space of the scattered elementary matter there must have been some one place where this primitive material had been most densely accumulated so as through the process of formation that was going on predominantly there, to have procured for the whole Universe a mass which might serve as its fulcrum .... The creation, or rather the development of nature, first begins at this centre and, constantly advancing, it gradually becomes extended into all the remoter regions, in order to fill up infinite space in the progress of eternity with worlds and systems .... At the primary stirring of nature, formation will have begun nearest this centre; and in advancing succession of time the more distant regions of space will have gradually formed worlds

and systems with a systematic constitution related to that centre .... There had mayhap flown past a series of millions of years and centuries, before the sphere of the formed nature in which we find ourselves, attained to the perfection which is now embodied in it; and perhaps as long a period will pass before Nature will take another step as far in chaos.[7]

When after a further lapse of time order is renewed in the region of the original order, we have the situation in which this ordered region is surrounded by a region where order has lapsed back into chaos, and this by a region where order is newly created, beyond which the original chaos still persists; and so it goes on. We seem therefore to be presented with a universe in which at any period of time some regions are oscillating between order and chaos, while in other regions the initial chaos has not yet been transformed by gravity into order.

> Millions and whole myriads of millions of centuries will flow on, during which always new worlds and systems of worlds will be formed after each other in the distant regions away from the centre of nature, and will attain to perfection. ... This infinity in the future succession of time, by which eternity is unexhausted, will entirely animate the whole range of space to which God is present, and will gradually put it into that regular order which is conformable to the excellence of His plan. And if we could embrace the whole of eternity with a bold grasp, so to speak, in one conception, we would also be able to see the whole of infinite space filled with systems of worlds and the creation all complete. But as, in fact, the remaining part of the succession of eternity is always infinite and that which has flowed is finite, the sphere of developed nature is always but an infinitely small part of that totality which has the seed of future worlds in itself, and which strives to evolve itself out of the crude state of chaos through longer or shorter periods.

How Kant could at the same time offer the reader an hierarchical universe in which moons orbit around planets and planets around suns, suns form galaxies and orbit around a massive luminous body (of which Sirius may be an example), and so on indefinitely – is not clear.[8] But here he has more in common with his contemporary J. H. Lambert (see Section B, chap. 4), who was still committed to the stable, clockwork universe of Leibniz, and who saw the hierarchy as permanent (and of finitely many steps). Wright, Kant and Lambert were all philosophical or theological speculators whose cosmological interests led them to attempt an explanation of the Milky Way. But the observational basis for their speculations was in each case vanishingly small. The last decades of the eighteenth century and the first of the nineteenth were to see the publication, in no less a journal than the *Philosophical transactions*, of equally speculative papers on 'the construction of the heavens'; but this time the author, William Herschel, was among the greatest telescope makers, and among the greatest observers, of all time. How his theories fared we shall see in Section C.

REFERENCES

1. I. Newton, *Optice* (London, 1706), Qu. 20. See ref. 59 of chap. 2 below.

2. William Whiston, *Praelectiones astronomicae* (Cambridge, 1707), Lect. IV; English trans., *Astronomical lectures* (2nd edn, London, 1728), 42.
3. William Derham, *Astro-theology* (London, 1715); cited from the 3rd edn (London, 1719), p. xvi.
4. *Ibid.*, p. xxxix.
5. *Ibid.*, 40, 52, 57-58.
6. On this see Simon Schaffer, "The phoenix of nature: Fire and evolutionary cosmology in Wright and Kant", *Journal for the history of astronomy,* ix (1978), 180-200.
7. Translation by W. Hastie, *Kant's cosmogony* (Glasgow, 1900), 142-5.
8. Some help with this is provided by Schaffer, *op. cit.*

2. Newton, Providence and the Universe of Stars

"Comme le Systême de Neuton paroît se contredire à l'égard des Fixes, qui, selon lui, se tirent les unes les autres, & demeurent pourtant immobiles, il faut commencer par éclaircir son sentiment, & faire voir qu'il n'implique aucune contradiction" (Voltaire, 1738).[1]

"A continual miracle is needed to prevent the Sun and the fixed stars from rushing together through gravity", wrote David Gregory in May 1694 in a memorandum of conversations with Newton.[2] Since gravity is a prime source of movement in the Newtonian universe, each star in it will be tugged by every other star and so will accelerate in the direction of the resultant of all these pulls. Yet among the stars actually visible in the sky there was, in the seventeenth century, no evidence of the slightest movement; on the contrary, observations since Antiquity confirmed that the 'fixed' stars were well-named, and apparently exempt from the effects of universal gravitational attraction.[3] In 1687 Newton, to judge by the first edition of the *Principia*, was oblivious of the problem; by 1694 he had reached the position he was to hold for the rest of his life. What had happened in the intervening years, and what form did God's providential action take?[4]

The writing of the Principia

Our story may begin in autumn 1685, when Newton had completed the substance of his *De mundi systemate* which was intended to conclude the *Principia* but which was not published until 1728, after Newton was dead.[5] In *De systemate* Newton uses an ingenious technique due to James Gregory, of which more anon, to prove that the stars lie at very great distances from the solar system;[6] indeed, by suppressing the work he delayed by nearly half a century the general appreciation of the scale of inter-stellar distances.[7] He concludes: "The fixed stars being, therefore, at such vast distances from one another, can neither attract each other perceptibly, nor be attracted by our Sun."[8]

A hostile critic might have commented, first, that nothing has been achieved with regard to the distances of the stars from each other, only proof of their distances from our local star which we call the Sun; and second, that the *sum* of many pulls (infinitely many?) may be large even if the individual pulls are imperceptibly small.

The following year, in drafting the *Principia* as published, Newton did not need any longer to banish the stars to the vast distances revealed by Gregory's method in order to sustain his claim that the pull of a star on the solar system (and vice versa) is in practice imperceptible. He now had a sensitive dynamical test for the presence of an external disturbing force on the solar system. For, by Proposition XLV of Book I, the planetary apsidal lines would then rotate, and this they do not do to an extent Newton regarded as significant. He sums this up in Proposition XIV of Book III ("The aphelions and nodes of the orbits of the planets are fixed"), and in corollaries[9] he remarks that this is because the stars are, and remain, at great distances:

Cor. 1. The fixed stars are immovable, seeing they keep the same position to the aphelions and nodes of the planets.

Cor. 2. And since these stars are liable to no sensible parallax from the annual motion of the Earth, they can have no force, because of their immense distance, to produce any sensible effect in our system.

Newton's correspondence with Bentley

It was late in 1692 that Newton's attention was drawn to questions of cosmology and cosmogony such as he had ignored in the *Principia*. That year Richard Bentley, the brilliant young chaplain to the Bishop of Worcester, had delivered the first series of sermons endowed by Robert Boyle, to be given annually on the evidences of Christianity.[10] On 7 November Bentley preached his seventh sermon, and set out to show that the observed structure of the universe could not have arisen out of an initial chaos by natural causes operating without the guiding hand of Providence: that if

> there were once no Sun no Starrs nor Earth nor Planets; but the Particles, that now constitute them, were diffused in the mundane Space in manner of a Chaos without any concretion and coalition; those dispersed Particles could never of themselves by any kind of Natural motion, whether call'd Fortuitous or Mechanical, have conven'd into this present or any other like Frame of Heaven and Earth.[11]

A theologian, Bentley was here moving into the province of natural philosophy, and before putting the sermon into print he wisely wrote directly to Newton for technical advice. Newton had immense sympathy with Bentley's enterprise and would himself on later occasions argue publicly along similar lines.[12] He therefore sent careful answers to Bentley's questions.

What, Bentley wished to know, would happen if matter were spread evenly throughout a finite space and then allowed to move freely under the action of gravity? It would, replied Newton, "fall down into ye middle of the whole space & there compose one great spherical mass. But if the matter was eavenly diffused through an infinite space, it would never convene into one mass but some of it convene into one mass & some into another so as to make an infinite number of great masses, scattered at great distances from one to another throughout all yt infinite space. And thus might ye Sun and Fixt stars be formed . . .".[13]

But, objected Bentley in his reply, if the matter was indeed evenly disposed throughout an infinite space—spread out with unbroken and absolute uniformity—surely every particle would remain motionless and in equilibrium, with no cause to move in one direction rather than another; so how could the Sun and fixed stars result? He was right, of course, to challenge Newton's response, but Newton gently explained how implausible and unreal such a perfectly uniform distribution was. To postulate a single particle exactly at rest in the middle of a finite mass was already asking too much:

> but that there should be a Central particle so accurately placed in ye middle as to be always equally attracted on all sides & thereby continue without

motion, seems to me a supposition fully as hard as to make ye sharpest needle stand upright on its point upon a looking glass. For if ye very mathematical center of ye central particle be not accurately in ye very mathematical center of ye attractive power of ye whole mass, ye particle will not be attracted equally on all sides.

And much harder it is to suppose that all ye particles in an infinite space should be so accurately poised one among another as to stand still in a perfect equilibrium. For I reccon this as hard as to make not one needle only but an infinite number of them (so many as there are particles in an infinite space) stand accurately poised upon their points. Yet I grant it possible, at least by a divine power; & if they were once so placed I agree with you that they would continue in that posture without motion for ever, unless put into new motion by the same power. When therefore I said that matter eavenly spread through all spaces would convene by its gravity into one or more great masses, I understand it of matter not resting in an accurate poise.[14]

Bentley also needed help with a paradoxical conclusion which seemed to him to apply in a universe with an infinite amount of matter, whether or not the matter is evenly distributed; namely, that every particle in such a universe rests in equilibrium because it is poised between pulls in opposite directions exerted, in every case, by infinite quantities of matter—such pulls being, in Bentley's view,[15] infinitely great and therefore equal as well as opposite!—so that the presence or absence of the (finite) Sun in one direction will make no difference to the motion of the Earth! Newton gently explained that not all infinites are equal: "And so a Mathematician will tell you that if a body stood *in equilibrio* between any two equal and contrary attracting infinite forces, & if to either of those forces you add any new finite attracting force: that new force how little so ever will destroy ye equilibrium & put ye body into ye same motion into which it would put it were those two contrary equal forces but finite or even none at all."[16]

Bentley of course had stumbled upon a genuine problem. Distributions of infinitely-many stars can easily be imagined in which the infinitely-many pulls have a *finite* sum and no theoretical difficulty results;[17] but in an infinite *and uniform* distribution of matter (or of stars), the resultant pull to one or other half of the sky is infinite (in modern terms: the gravitational potential is undefined). Discussion of the profound difficulties in comprehending such a universe was to be resumed two centuries later.[18] Newton has already hinted at his own approach: he will pair off pulls in one direction with equal and opposite pulls in the other and for dynamical purposes will deem the pairs of pulls, and the matter causing them, effectively not to exist—whether the pairs are finite *or infinite* in number.

Newton's reward for this reply and a further letter sent by way of postscript was a manifesto from Bentley of enormous length,[19] summarizing with commentary the whole argument of his seventh sermon. The sermon is mainly concerned—as is their correspondence—with the solar system and in particular the explanation of the pattern of planetary and cometary orbits. But Bentley does rehearse, with great effect, the position the two have reached regarding the stars. First, they are agreed that the stars do *not* form a finite system, with each star in orbit around the centre of the system like the planets round the Sun. Such a system would of course be both finite and stable (and indeed our Galaxy is just such a system, though lacking a single massive central body); but so fixed, so stationary had the stars become in European consciousness that this possibility could be discounted without further argument:

> I grant yt if ye whole World was but one Sun and all ye rest planets moving about him, they would not convene. But in several fixt stars, yt have no motion about each other; they with their systems of planets would all convene in ye common center of mundane gravity; if ye present world was not susteind by a divine power.
>
> Sir, In a finite world where there are *outward* fixt starrs, this seems plainly necessary. But in ye supposition of an infinite space, let me ask your opinion. I acquiesce in your authority, yt in matter diffused in an infinite space, tis as hard to keep those infinite particles fixt at an equilibrium, as poise infinite needles on their points upon an infinite speculum. Instead of particles, let me assume Fixt starrs or great Fixt Masses of opake matter; is it not as hard, yt infinite such Masses in an infinite space should maintain an equilibrium, and not convene together? so yt though our System was infinite, it could not be preserved but by ye power of God.[20]

This devastating argument, virtually taught to Bentley by Newton himself, was to dominate Newton's own subsequent attempts to reconcile the fixity of the stars with the action of universal gravity. And it was repeated for all to read in the published version of Bentley's sermon:

> yet, we say, the continuance of this Frame and Order for so long a duration as the known ages of the World must necessarily infer the Existence of God. For though the Universe was Infinite, the Fixt Starrs could not be fixed, but would naturally convene together, and confound System with System: for, all mutually attracting, every one would move whither it was most powerfully drawn. This, they may say, is indubitable in the case of a Finite World, where some Systems must needs be Outmost, and therefore be drawn toward the Middle: but when Infinite Systems succeed one another through an Infinite Space, and none is either inward or outward; may not all the Systems be situated in an accurate Poise; and, because equally attracted on all sides, remain fixed and unmoved? But to this we reply; That unless the very mathematical Center of Gravity of every System be placed and fixed in the very mathematical Center of the Attractive Power of all the rest; they cannot be evenly attracted on all sides, but must preponderate some way or other. Now he that considers, what a mathematical Center is, and that Quantity is infinitly divisible; will never be persuaded, that such an Universal Equilibrium arising from the coincidence of Infinite Centers can naturally be acquired or maintain'd.[21]

But for the present, Newton kept his own counsel, and in his answer to Bentley's letter ignored the problem.

Drafts for a second edition of the Principia

The correspondence with Bentley was taking place about the time that

Newton was preparing for a second edition of the *Principia*. Such plans at this time proved abortive, but substantial drafts survive in chaotic disorder among the Newton manuscripts in the University Library, Cambridge. I examined these pages some years ago in the hope that Newton would have been provoked by Bentley's cross-examination to develop his thinking on the universe of stars, and was able to piece together what proved to be a series of drafts[22] for a new Theorem XV of Book III, to replace the brief corollaries to Theorem XIV. In them Newton develops his thinking on the role of the stars more completely than anywhere else known.

The confusion in the statement and first draft of the theorem bears witness to its duality of purpose. The theorem is to replace the corollaries, which set out how the individual stars affect the solar system; but it is also, and primarily, to make some answer to Bentley's dilemma over the nature of the stellar system as a whole and the dynamical influence of the system on its individual members:

> Proposition XV. Theorem XV. The fixed stars are at rest in the heavens and are separated by enormous distances from our Sun and from each other.[23]

Newton first argues briefly that the stars are seen to be at rest in relation to each other, and that the absence of observed annual parallax proves they are very distant from the solar system, so distant that for them to orbit about the solar system (as the planets about the Sun) is dynamically impossible. They are therefore at rest, not merely relatively, but absolutely.[24]

This may satisfactorily prove that the stars are at rest and at substantial distances from us. But it does nothing to prove they are at similarly-great distances from each other, as implied in the theorem; still less does it answer Bentley's worries as to how the system of stars, whether finite or infinite, can be dynamically in equilibrium. But for the moment Newton is preoccupied with strengthening the existing argument, by showing that the stars are actually very much further than the minimum distance which is all that can be inferred from the astronomers' failure to detect annual parallax.[25] He will do this, as in *De systemate*, by exploiting a brilliant technique due to James Gregory; Newton owned a copy of his rare *Geometriae pars universalis* (1668) in which the technique is set out.[26] The details need not detain us here, but the principle is important. One assumes that all stars, including the Sun, are equally luminous, so that the fainter a star appears, the greater its distance from us. One further assumes that nothing physically untoward happens to starlight on its journey through space, so that the apparent brightness falls off simply with the square of the distance. It follows that the ratio of the apparent brightnesses of two stars can readily be converted into the ratio of their distances. In particular, if one of the stars is the Sun and if the Sun's distance is known, then the ratio can, so to speak, be calibrated to give the *actual distance* of the other star.

The problem is a technical one: how actually to carry out the photometric comparisons—how to determine the number of stars like (say) Sirius needed to equal the Sun in brightness. Gregory solved this problem by using a planet as an intermediary between the Sun and Sirius. One waits until Saturn, for example, appears as bright as Sirius. The problem of comparing the Sun's light with that of Sirius is then the same as that of comparing the Sun's light with that of

Saturn; or, better, of comparing the light that comes to us directly from the Sun with the light that comes to us from the Sun via Saturn. This last comparison depends upon distances and other data internal to the solar system, which we regard as fully understood. Gregory himself realised that the value he had for the distance of the Earth from the Sun (the *astronomical unit*) was much too small and that as a result his distance for Sirius of 83,190 a.u. was also too small. Newton was better informed, and extends the draft of the present theorem to get for Sirius a distance "in round numbers" of one million a.u.—a value actually in excess of the true figure.[27]

Gregory's method depends, as we have seen, on several assumptions; in particular, that the apparent brightness falls off with the square of the distance, and that the reflectivity of Saturn is correctly known. Newton concludes this first draft by justifying these assumptions.[28] He points out that a given error in the reflectivity will affect his distances by a factor of only the square root of the error, "and I neglect trivialities of this kind".[29] As to possible loss of light from a star as it journeys through space, he points out (as in *De systemate*) that a given loss over the distance from one star to the next will be compounded over every similar stage of the journey, in which case all stars except the nearest will be reduced to invisibility if the given loss is at all significant. He believes the general appearance of the night sky shows this not to be the case.

Newton now realizes that this draft is not adequate since he is committed to a theorem about the spatial distribution of the stars one from another, and so he prepares (see Figure 1) an interpolation to meet this commitment:

> But since the heavens give free passage and the bodies of the Sun and stars gravitate towards each other in proportion to the distances yet do not fall towards each other or acquire any perceptible motion as a result of that gravity, it must be that the stars are separated from each other as they are

FIG. 1 (opposite). ULC Add. MS 3965, f280r (courtesy Cambridge University Library), showing amendments to Newton's first attempt to draft Proposition XV. The first two words ("tangentibus orbium") conclude the first paragraph. Newton then decides to jot down some calculations involving the apparent diameter of Saturn and its ring (27″), and having disfigured this sheet he writes the two words again on the facing page and continues there.

He later realizes that he must discuss the spatial distribution of the stars one from another, so he returns to this sheet and poses the problem ("Cum autem coela . . . insensibilis evadat"—note how "prope" is changed first to "omnino" and then to the innocuous "plane") and indicates that he will solve it by examining stars of successive apparent magnitudes ("Fixarum quidem . . . colligere licet"). The first sentence is recognizable in the final draft ("Cum autem corpora . . .", see Appendix).

Newton then claims a match between predictions from the model and data in the star catalogues ("Sunt tridecim . . . respective"—note the gaps left for the actual numbers to be inserted later), but this he immediately redrafts ("Nam si . . . nondum referuntur") and will again redraft elsewhere (see Figure 2).

The sheet concludes with three passages that survive to be recognizable in the final draft: a hint that the telescopic stars seem as numerous as the model requires ("Adhibitis vero Telescopijs . . . videatur"); a concession to allow some variety among the stars ("Stellarum tamen distantias . . . et contra"); and a justification of the assumption that the Sun is a star ("Cum fixae omnes . . . habenda sunt"). Two sentences on annual parallax are keyed in ("Cum nulla observetur . . . vicibus & amplius") but will disappear at the next stage.

Just below the Gregory calculations at the top of the page, Newton has left a small space which he will later use to jot down the numbers 1. 27. 125 . . . and 1. 13. 49 . . . squeezed out of our Figure 2.

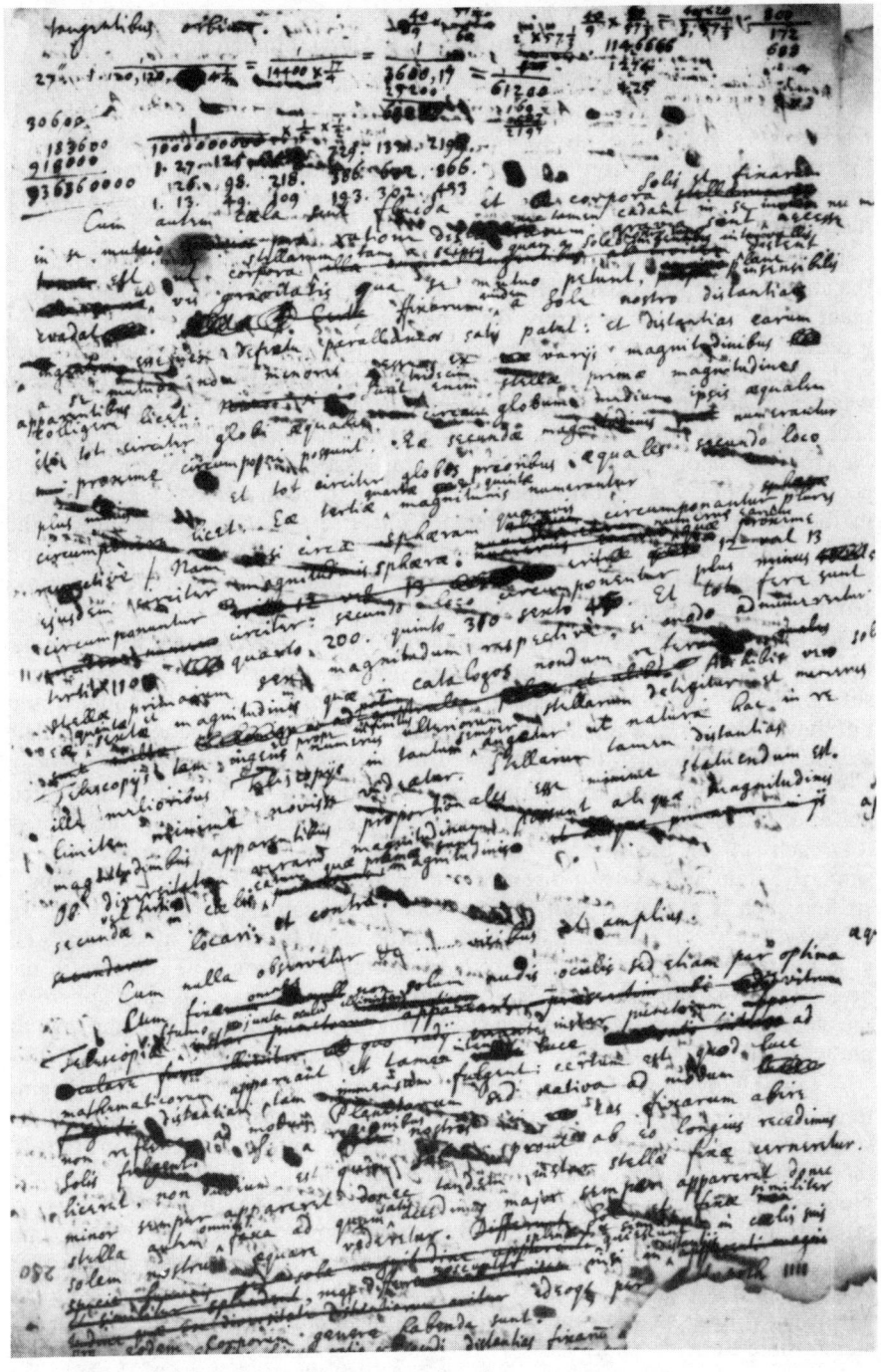

from the Sun, by such huge intervals that the force of gravity by which they pull each other proves to be nearly [*prope*] imperceptible.[30]

Newton is not happy with *prope*, perhaps because it implies that the pull is nevertheless perceptible; he alters it to "completely" (*omnino*), but evidently feels this is going too far to the other extreme, and settles for the vague "plainly" (*plane*). Newton is reluctant to face the dilemma that, however small the individual pulls, if they are non-zero unpleasant possibilities arise when the number of stars available to do the pulling is infinite.

But what can Newton possibly do to justify his claim about the distances of the stars one from another? His unspoken objective is to give a plausible argument to show that the *real* universe is not very different from an idealized one, a geometrical model in which the stars are distributed throughout infinite space with exact regularity—so that each such star is symmetrically poised between *equal and opposite* gravitational pulls and therefore remains at rest. It was just such a universe that Bentley postulated—an infinitely "hard" supposition, Newton had said, "yet I grant it possible, at least by a divine power". But this geometrical model is almost impossible to relate to observations by astronomers in the real universe. A more convenient universe as far as the testing of the model against observation is concerned would be one in which each star—and the Sun in particular—is surrounded by stars of which the nearest are all at (say) 1 unit of distance, the next nearest all at 2 units, the next at 3 units, and so on.

An objector might say that we could arrange this for the Sun, for example, but that it would be geometrically impossible to arrange for *every* star to be surrounded by other stars at *exactly* these distances; further, the model would not have the perfect regularity throughout which seems the only complete escape from the accelerations generated by universal gravitation.

We can go some way to meeting these objections if we have the stars which are at 1 unit of distance arranged (on the sphere of unit radius) in such a way that each star on the sphere is at least 1 unit from the others on the sphere; similarly, stars at 2 units of distance will be arranged on the sphere of radius 2 so that each is at least 1 unit from others on the same sphere; and so on. In this way, every star will be at 1 unit or a little more from *some* other stars, and at a greater distance from all the remaining stars—as would be the case in the perfectly regular model. Further, if we crowd as many stars on each sphere as the geometrical constraints permit, we shall come closer still to the regular model.

A valuable dynamic property of our Sun-centred nest of spheres now begins to emerge. For a sphere of large radius n units, the numerous stars on it will be spread out in a good approximation to spherically uniform distribution; that is, these stars taken together will approximate dynamically to a thin uniform shell. Now such a shell, by *Principia*, Book I, Proposition LXX, has zero net gravitational attraction, not only on the Sun at the centre, but on *any* of the stars inside the shell. In other words, as we travel in imagination further and further from the Sun in our model, the stars at these distances, though huge in number, nevertheless have slighter and slighter dynamical effect on the Sun and on the stars visible to us.

How many stars can we get on the unit sphere such that the stars are separated by distances of at least one radius? Or, equivalently, how many spheres equal to a given sphere can be packed so as each to touch the given sphere? Kepler asserted that the number is at most 12,[31] but Newton thought it might be 13. Now on the sphere of radius 2 units we have four times the surface area available and so we might expect to cram on it four times as many stars each separated from the others by at least 1 unit; that is, about $4 \times 12\frac{1}{2} = 50$. Similarly, on the sphere of radius 3 units we have nine times the surface area and so room for nine times 12 or 13; and so on.

Now these predictions from the geometrical model are couched in terms of *distances*. To test them against observations we must relate them to the information in the star catalogues; that is, to the lists of star positions and *apparent brightnesses*. Obviously we must assume once more that the nearest stars are the brightest (first magnitude), for how else can we hope to test theory against observation; and we note with satisfaction that the number of first-magnitude stars is customarily put very near to our 12 or 13 derived from the geometrical model. But what corresponds in the sky to the stars at 2 units of distance? And at 3, 4 and so on? Newton, like William Herschel a century later, assumes the answer to be: stars of magnitude 2, 3, 4 and so on. Confident of both this detailed assumption and the general correctness of his model, Newton continues (see Figure 1):

> That the stars are at huge distances from our Sun is clear enough from the absence of parallax; and that they lie at no less distances from each other may be inferred from their differing apparent magnitudes. For there are 13 stars of the first magnitude and roughly the same number of equal spheres can be arranged around a central sphere equal to them. The stars of the second magnitude number about [he does not have a star catalogue to hand, but he is confident of a successful outcome and so leaves a gap to be filled in later] and this is roughly the number of spheres, equal to the previous ones, which can be arranged around them in the second position. Those of the third, fourth and fifth magnitudes number   respectively. . . .

Newton feels that the exposition is not going as well as it might, so he retains the first sentence from the above quotation and tries again with his statement of the comparison:

> For if around some sphere there are arranged more spheres of about the same size, the number of spheres which surround it closely will be 12 or 13; at the second stage about 50; at the third, about 110 [roughly 9×12]; at the fourth, 200 [$16 \times 12\frac{1}{2}$]; at the fifth, 310 [about $25 \times 12\frac{1}{2}$]; at the sixth 450 [$36 \times 12\frac{1}{2}$]. And this is roughly the number of stars of the first six magnitudes respectively, provided you reckon in those stars of the fifth and sixth magnitudes which are not yet catalogued.

Completely confident that the predictions from the model will make a satisfactory match with the observational data for the naked-eye stars, Newton turns his attention to the telescopic stars. Every improvement in the telescope reveals more stars, so that in this regard "nature seems to know no bounds" and the evidence, such as it is, confirms the limitless stars of the model. At this point Newton warns the reader that he does not insist rigidly on the brighter stars being invariably nearer, for in reality there is some variety among the stars, and stars of the second or even third magnitude may be located in the regions of the first magnitude and vice versa. Finally, he completes this first revision of his theorem by adding a paragraph in justification of Gregory's assumption that the Sun is indeed a typical star.

Perhaps immediately, perhaps after a lapse of time, Newton takes a fresh sheet of paper[32] and embarks on a further revision of his draft Proposition. He copies out the statement and first paragraph with little change; but then Newton strikes out this paragraph and begins afresh. No doubt he felt a serious lack of balance in the existing draft, in which the great distances of the stars from the solar system are first assumed and only afterwards proved. Now he *begins* by using the absence of observed annual parallax to prove the great distances, and *then* he shows how this implies that the stars are at rest not only relatively but absolutely.

He then faces up to the problem of the distances of the stars from each other ("Ad ingentes autem distantias tam a seinvicem quam a Sole per universa coela spargi sic etiam colligo"), and copies out the previous discussion of the geometrical model, continuing beyond it to the preamble to the Gregory argument.[33]

Here or hereabouts he becomes dissatisfied with the obvious incompatibility of the nests of concentric spheres which comprise the model actually studied (and which taken literally would imply a universe with the solar system as its focus) and the claim of the Proposition (which embraces the stars in general). Anyone who has expounded this concept to an audience will sympathize with Newton in his predicament. He attempts to escape from it by setting out on a new sheet of paper[34] (see Figure 2) a revised model in which the points (stars) are spread out at roughly equal distances and only then are, so to speak, assigned each to the nearest of the nest of concentric spheres:

> If we imagine six concentric spheres drawn with radii in the arithmetic proportions 1, 2, 3, 4, 5 and 6, and lucid points spread out at roughly equal distances throughout the space; and if the points are counted which are nearest to the individual spheres; then the number of those nearest to the first sphere will be to the number of those nearest to the second in about the ratio of the surfaces, that is, as the squares of the radii 1, 2, namely 1:4. And similarly the number of those near to the second surface will be to the numbers of those near to the third, fourth, fifth and sixth as the ratios of the surfaces; that is, as the square of the radius 2—namely, 4—to

FIG. 2 (opposite). ULC Add. MS 3965, f 74r (courtesy Cambridge University Library). Further attempts by Newton to express the relationship between predictions from his geometrical model and data in the star catalogues. For Newton's argument at this point, see our main text, but note how the first paragraph dies away as Newton at last consults the observational data and finds an unwelcome clash with the predictions: he immediately deletes all references to fifth and sixth magnitudes before going on to experiment at the foot of the page with more complex relationships between predictions and data. The figures 1. 27. . . . partly burned away are jotted down again by Newton on f 280r (Figure 1); he is listing the spaces inside spheres of radii 1, 3, 5, . . . and deriving the spaces outside one sphere but inside the next, hoping these will be proportional to the numbers of stars of successive magnitudes.

In the second paragraph, Newton comes close to a realization of 'Olbers's Paradox' when he notes that at two units of distance the stars are four times fainter and four times more numerous ("quadruplo obscuriora et quadruplo plura"), and similarly for greater distances.

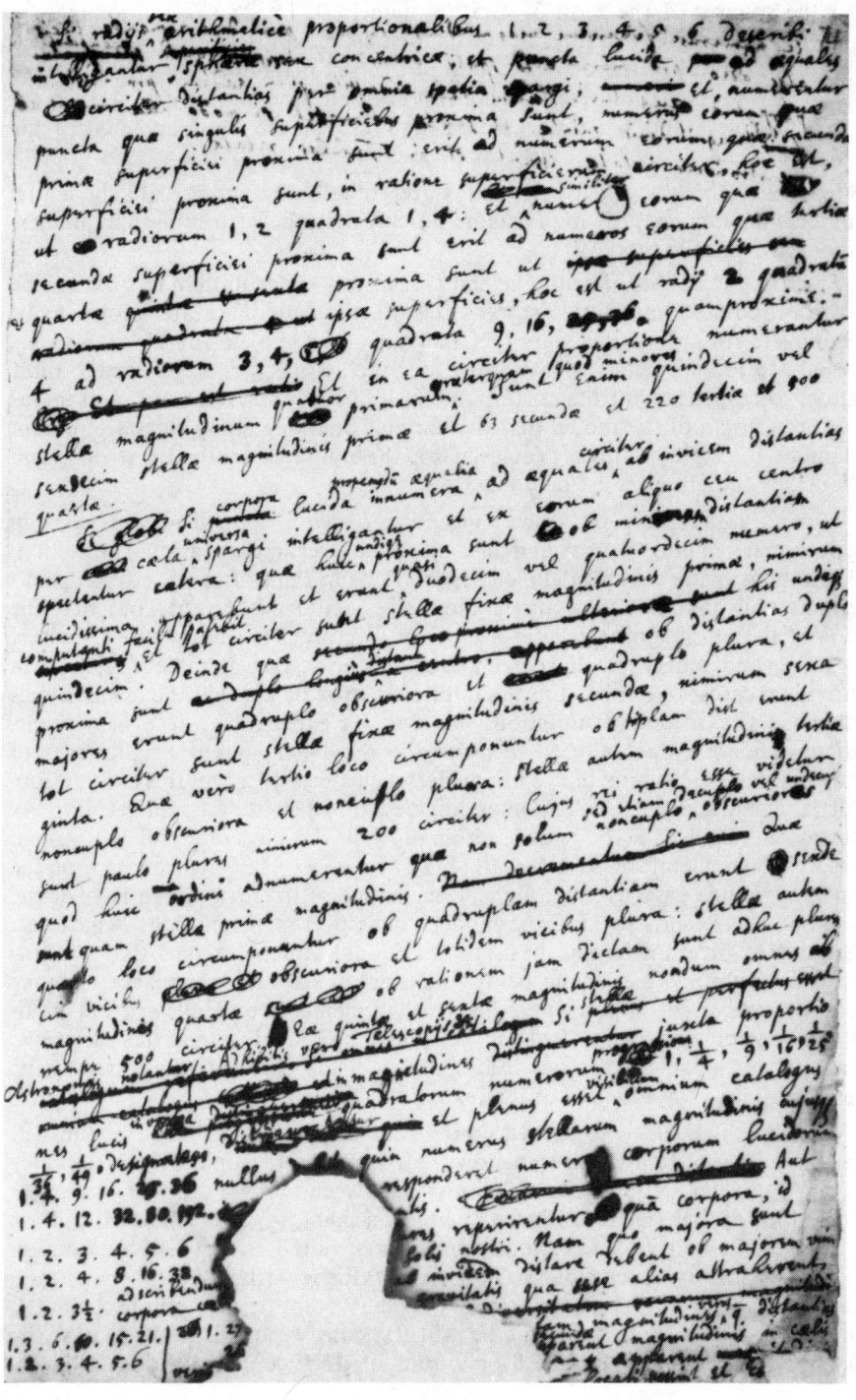

the squares of the radii 3, 4, 5, 6—namely 9, 16, 25, 36, approximately. And the numbers of the stars of the first six magnitudes are roughly in this proportion. For there are 15 or 16 of the first magnitude [for the first time in this investigation, Newton actually looks up the numbers of stars in the catalogues] and 63 of the second and 220 of the third and 500 of the fourth. . . .

Newton here breaks off, for the ratios are going badly wrong, and the actual numbers of stars are growing far more quickly than the corresponding numbers predicted from the model![35]

To escape from the difficulty, he first crosses out all mention in this paragraph of fifth- and sixth-magnitude stars—after all, the catalogues are known to be incomplete for the fainter stars (yet they are *too* complete for Newton's purposes!). In this way he is left with just four predictions (or, more exactly, three ratios) to compare with the corresponding totals of observed stars; and it was a basic strength of the model that the familiar first-magnitude stars are present in about the right number. Yet even so, the numbers are badly wrong, and about this something must be done.

As well as adopting his basic conception of a universe filled with stars in approximately uniform array and all subject to universal gravitational attraction, Newton has made various detailed assumptions in the course of setting up this test, notably that the stars are in themselves uniformly bright, and that the stars of the nth magnitude lie at n units of distance. He cannot *prove* the first of these, and indeed we have seen that he allows considerable deviation in individual cases; but unless this assumption is broadly correct, the test becomes vacuous. The second assumption, however, is expendable; it is simply his technical inability actually to *measure* the relative apparent brightnesses of stars that has forced him, like William Herschel later,[36] to adopt this assumption. But in fact the ancient astronomers divided the stars they could see into six classes or magnitudes for reasons originating in the physiology of the human eye. It would therefore be a remarkable coincidence if 36 typical stars of the sixth magnitude gave as much light to us as one of the first, which is what the assumption amounts to; and the modern definition of magnitudes, which was designed to give a scientific basis to traditional values, implies that a sixth-magnitude star is, other things being equal, not at six but *ten* times the distance of a first-magnitude star, a figure brilliantly anticipated by Halley in 1721.[37] In short, Newton can escape from his difficulty by claiming that the successive magnitudes extend further out into space than the corresponding predictions and that this is why the sets of figures appear inconsistent.

Once having decided to accord himself this degree of freedom, Newton will be able to complete his task to his own satisfaction. He would of course prefer to find an actual mathematical relationship between the two sets of figures, and at the foot of the page (see Figure 2) and in a blank space on the earlier draft (see Figure 1) he tries out various sequences of numbers, but without coming to any satisfactory conclusion. So he settles for the remark that third-magnitude stars correspond to distances not only of 3 units (that is, $\sqrt{9}$) but also of $\sqrt{10}$ and $\sqrt{11}$; and in the final draft he will for consistency also add that sixth-magnitude stars correspond to 8 or 9 units of distance, which brings Newton very close to the correct relationship between first and sixth magnitudes. But this is the result of luck rather than judgement.[38]

Newton is now within sight of his final draft exposition of this awkward argument:

> If numberless and roughly equal shining bodies are thought of as scattered through all the heavens at roughly equal distances from each other, and if from some one of them, as the centre, the rest are observed: those nearest to this one on every side, being at the least distance, will appear brightest and will total about 12 or 14, as readily appears from calculation. And the stars of the first magnitude are of about this number, namely 15.
>
> Next, those stars which are nearest to these on every side, being at twice the distance, will be four times fainter and four times more numerous; and of about this number are the stars of the second magnitude, namely 60.

To modern eyes, this remark that the stars at two units of distance are four times more numerous and four times fainter brings irresistably to mind the familiar exposition of the so-called Olbers's Paradox: with increased distance the reduction in light from each individual star in an infinite and regular star system is offset by the increased numbers of stars, so that the *total* light from the stars at two units is the same as that from the stars at one unit, and the same as the total light from the stars at three, four, . . . units. It can easily be shown that in consequence the whole sky in an infinite, homogeneous universe would be ablaze with the equivalent of sunlight, which is manifestly not the case. This grave threat to Newton's model of the universe is overlooked by him, obvious though it seems to modern eyes; no doubt his mind was preoccupied with other threats to his model. But in 1721 he was to be in the Chair at a meeting of the Royal Society when Halley gave the first known exposition[39] of the 'paradox', "which I have heard urged" (by David Gregory, I would guess). Did Newton choose to ignore the threat, or was he convinced by Halley's fallacious "resolution" of the difficulty—or was he simply asleep?

Newton's continuation of his argument makes his failure to perceive the 'paradox' still more surprising:

> Those stars distributed round about at the third stage, being at three times the distance, will be nine times fainter and nine times more numerous; but the stars of the third magnitude are somewhat greater in number, namely about 200. The reason for this seems to be that in this figure are included not only those stars nine times fainter than those of the first magnitude but also those ten or eleven times fainter. Those distributed round at the fourth stage, being at four times the distance, will be sixteen times fainter and the same number of times more numerous; but, for the reason just given, the stars of the fourth magnitude are more numerous, namely about 500.

Catalogues of the fifth and sixth magnitudes, he adds, are not yet complete, and the telescopic stars he dismisses as in the earlier draft. He then makes two significant comments. First, he says that were the star catalogues complete and were successive magnitudes to be *defined* to correspond to the inverse square of the distances (as he had at first believed to be the case), there would be no doubt (*nullus dubito*) that the numbers would correspond. Second, he adds the qualification that if the numbers of observed stars are (consistently) too large, "that should be ascribed to the size of our Sun. For the greater the heavenly bodies are, the further they ought [*debent*] to be distanced from the others because of the greater gravitational force by which they would pull the others to them".[40]

The problems are now resolved. Newton interpolates the gist of this discussion into the appropriate place in his draft,[41] and transcribes the Gregory argument and the related supporting considerations.[42] The chief novelty at this stage is a more extended investigation of the size of the Sun itself: the number of first-magnitude stars, being greater than expected, suggests that the Sun is above the average star in size in the ratio of something like 6 to 5. The Latin text of the complete Proposition is transcribed in the Appendix.

Related Newton documents

Among the Newton manuscripts in Cambridge University Library are three rough drafts of expositions of elementary astronomy, one or more possibly for use in undergraduate lectures.[43] They are undated but, to judge from the handwriting, may belong to the early 1690s. One is in Latin and entitled "Machinae Mundanae Descriptio brevis",[44] and is evidently a modification of an earlier version of the same. The first paragraph declares that the Sun and the stars shine by their own light and are bodies of the same kind scattered throughout all the heavens and maintain steadily (*constanter*) their given relative positions. The Sun, if removed further and further from us, would appear in turn as of the first, second, . . . magnitude and would then vanish from sight; similarly, a star brought close to us would appear like the Sun. He continues: "Stellae fixae haud minoribus ab invicem intervallis distant quam a sole nostro. Id ex eorum magnitudinibus colligitur. Nam si . . .", where the paragraph breaks off. This is strongly reminiscent, in phraseology as well as argument, of f280^r discussed above: ". . . distantias earum [fixarum] a se mutuo non minores esse ex variis magnitudinibus colligere licet. Nam si . . .", and it may well be that the "Descriptio" was written with f280^r or a related draft before him.

The other two documents were transcribed by A. R. and M. B. Hall in their volume of *Unpublished scientific papers of Isaac Newton*.[45] The first is headed "Cosmography. Chap. 1. Of the Sun & fixt Stars",[46] and Newton begins by rehearsing the basic properties of the Sun and stars, along the lines of the "Descriptio". He then explains the assumption of equal luminosity from which one can estimate the distances of first-magnitude stars, and makes to state the result, though leaving the actual quantities to be filled in later. He continues:

> Yet this is to be understood wth some liberty of recconning. For we are not to account all the fixt starrs exactly equal to one another, nor placed at distances exactly equal nor all regions of the heavens equally replenished with them.
>
> For some parts of the heavens are more replenished wth fixt stars yn others, as the Constellation of Orion with greater or nearer stars & the milky way with smaller or remoter ones. For ye milky way being viewed through a good Telescope appears very full of very small fixt stars & is nothing else then ye confused light of those stars. And so ye fixt clouds & cloudy stars are nothing else then heaps of stars so small & close together

that without a Telescope they are not seen appart, but appear blended together like a cloud.

This is the only document known to me where Newton explicitly recognizes that nebulae (which he identifies with star clusters) and the Milky Way are visible proof of how far the real universe departs from his perfectly regular model. However, he embarks on the familiar argument for a regular distribution of the naked-eye stars:

> Were all the fixt stars equal & placed at equal distances from one another, the number of the stars next about us would be 12 or 13, those next about them 50, those next about them 110, those next about them 200, those next about them 300, those next about them 450 or thereabouts. And accordingly there are about 12 or 15 stars of ye first magnitude, about 50 or 60 of ye second, about 200 or

Here he breaks off, realizing, as in the draft theorem, that the sets of numbers do not tally, and he deletes the last and incomplete sentence. He tries again, and a third time, before settling at the fourth attempt on the weak conclusion: "And tho their magnitudes & distances be not equal yet this affords ye true reason why the smallest stars are the most numerous. . . ."

The third document[47] of the trio is headed "Phaenomena", and it is notable for Newton's discovery of a mathematical relationship between the number of stars of successive magnitudes, such as he had looked for in vain when drafting the theorem. In "Phaenom. 6" he yet again is surprised to find the sets of numbers do not tally, but this time he amends his draft to read:

> But if the Lamps [which for didactic purposes represent stars] be distinguished into magnitudes according to the quantity of their light decreasing in a geometrick progression (wch is the case for the fixt stars) there may be 120 or 140 of the third magnitude, 3 or 4 hundred of the fourth nine hundred or a thousand of the fift.
>
> And so it is in the fixt stars. There are about 12 or 14 stars of the first magnitude, about 50 of the second magnitude about 160 of the third about 4 or 5 hundred of the fourth & so on. For the stars of the smaller magnitudes being distinguished by their light decreasing in a geometric proportion, there are reccomed about three times as many stars of the third magnitude as of the second & about three times as many of the fourth magnitude as of the third, & so on.

To modern eyes this passage is truly remarkable. When he says of successive magnitudes that the brightnesses decrease "in a geometric proportion", Newton is anticipating the nineteenth-century application of Fechner's Law which, in Pannekoek's words, "implies that between successive magnitudes there is a constant ratio, which must be derived from photometric measurements". Indeed, if we replace Newton's "about three times" by 3·98, we have exactly the modern definition of magnitudes![48]

The dissemination of Newton's world picture

When David Gregory visited Newton in May 1694, the essentials of Newton's world picture were established. Newton explained to Gregory his uncle's method of determining the distances of the stars; he explained to him the model of a uniform distribution of stars, and how it might be related to observations, remembering "other inequalities can compensate"; he probably indicated the fundamental role of Proposition LXX of Book I of the *Principia*, concerning gravity within a uniform shell; and he advised him that "a continual miracle is needed to prevent the Sun and the fixed stars from rushing together through gravity".[49]

Gregory revealed something of this in his popular *Elementa* (1702). In Proposition XXI, "To explain the several Orders of the Fix'd Stars, into which they are divided on account of their differing apparent Magnitude; and to give an account of this difference", he argues that those who "make their [the stars'] different distance to be the cause of their different Magnitude" are supported by the closeness of the numbers of first- and second-magnitude stars to 13 and 4×13; but thereafter "the Matter does not go on so well, (which made *Kepler* of another Opinion,) as is evident even from hence, that upon the first cast of our Eyes upon the Heavens, some Tracts of the Firmament appear fill'd with innumerable Fix'd Stars, whereas others are found to be almost empty and void of any".[50] Gregory gives no sign of concern at this large-scale divergence from a uniformly-distributed star system. Perhaps he was simply content to leave the whole problem of sustaining the universe in God's hands, since he and Newton had agreed that God was unquestionably involved.

In his textbook, however, he seems to suggest that *because* the system is boundless it automatically follows that there is no danger of gravitational collapse: "The indefinite Number of those Systems, included in no Space, is the Reason why they don't run into one, but being separated from one another, will for ever stand in the Universe, as Marks of the Power and Wisdom of their Almighty Creator."[51] This seems to have been the opinion of Edmond Halley, when discussing "the infinity of the sphere of the fix'd stars" before the Royal Society in 1721 with Newton in the Chair. In the text as printed he says: "But if the whole [System of the World, that is, of the stars] be Infinite, all the parts of it would be nearly *in aequilibrio*, and consequently each fixt Star, being drawn by contrary Powers, would keep its place; or move, till such time, as from such an *aequilibrium*, it found its resting place."[52] This may suggest that Halley thought the proper motions of three stars which he had recently discovered,[53] and which showed at long last that the stars were *not* 'fixed' after all, were oscillations about an equilibrium position; whereas, of course, the movement of a star will increase the gravitational pulls in the direction of the continued movement and decrease those in the opposite direction, so that such oscillations are impossible.[54]

As for Newton, however frank he may have been in conversation with Gregory, his own printed references to his conception of the universe of stars were fragmentary and sparing. That God is involved is implied in a Query added to the Latin edition (1706) of the *Opticks*, where Newton asks:

> To what end are comets, and whence is it that planets move all one and the same way in orbs concentric, while comets move all manner of ways in orbs very eccentric, and what hinders the fix'd stars from falling upon one another?[55]

In 1713, when a second edition of the *Principia* at last appeared, Newton had been preoccupied with public affairs in London for the better part of two decades, and the new version was edited by Roger Cotes of Trinity College at the instigation of Bentley, by then Master of the college. In consequence, readers were not given sight of the elaborate Proposition XV drafted in the 1690s; instead, they had to make what they could of a succinct addition to Corollary 2 of Proposition XIV: "Not to mention that the fixed stars, everywhere promiscuously dispersed in the heavens, by their contrary attractions destroy their mutual actions, by Prop. LXX, Book I."[56]

The same edition concludes with the first version of the famous General Scholium, in which Newton draws attention to the regularities in the solar system, declaring: "This most beautiful system of the Sun, planets and comets, could only proceed from the counsel and dominion of an intelligent and powerful Being." In drafts of the General Scholium he also remarks on the providential arrangement of the fixed stars;[57] but this thought did not survive into the printed text of 1713, and it was only in the third edition of 1726 that Newton added the comment: ". . . and lest the systems of the fixed stars should, by their gravity, fall on each other, he hath placed those systems at immense distances from one another."[58] And with such clues his readers had to be content.

Newton, God and the stars

What, then, appears to be Newton's conception of the universe of stars? First, it is clear that gravity operates throughout the universe, and that a 'miracle' of divine intervention is necessary to prevent gravitational collapse. God does not simply *countermand* gravity permanently and totally, for then a finite star system would be possible; nor is it sufficient for God to leave great distances between stars, for this he has done and a miracle is still necessary. Rather, the system of the stars, like the system of planets and comets, has been constructed with providential ingenuity to minimize the tendency to gravitational collapse. In particular, the Sun can be regarded as surrounded by other stars in such a way that the stars at a given distance taken together are dynamically similar to a uniform shell, and so exercise at most a slight resultant gravitational force on the Sun *and* on the stars nearer to the Sun than the given distance.

Slight, but not always zero. Just as the solar system suffers from perturbations "which will be apt to increase, till this system wants a reformation",[59] so the structure of the stellar system is in danger in the long term. But there are important differences. The solar system is finite and there is an evident and striking pattern in the motions of the primary and secondary planets; as Newton told Bentley:

> To make this systeme therefore wth all its motions, required a Cause wch understood & compared together the quantities of matter in ye several bodies of ye Sun & Planets & ye gravitating powers resulting from thence, the several distances of the primary Planets from ye Sun & secondary ones from Saturn Jupiter & ye earth, & ye velocities wth wch these Planets could revolve at those distances about those quantities of matter in ye central bodies. And to compare & adjust all these things together in so

great a variety of bodies argues that cause to be not blind & fortuitous, but very well skilled in Mechanicks & Geometry.[60]

By contrast with the solar system, the system of the stars is infinite and at rest and the providential pattern in its spatial ordering far from obvious—indeed it has required a major effort by Newton to uncover it. But the need for providential intervention to *preserve* the system in good order is to Newton very evident, since God has not in fact chosen to give it the perfect symmetry that is "as hard as to make not one needle only but an infinite number of them . . . stand accurately poised upon their points".

Newton was provoked by Bentley to examine the system of the stars, and he uncovered proof of both providential skill in the ordering of the system and the need for providential intervention in its maintenance. As Samuel Clarke wrote in criticism of Leibniz's conception of God: "The notion of the world's being a great machine, going on without the interposition of God, as a clock continues to go without the assistance of a clockmaker, is the notion of materialism and fate, and tends . . . to exclude providence and God's government in reality out of the world."[61]

Acknowledgements

I am indebted to my colleague Dr D. T. Whiteside for help readily given in innumerable conversations over the years; to the referees who commented without the customary protection of anonymity *vis-à-vis* the author; and to the Librarians of the Royal Society and Cambridge University Library for permission to use manuscripts in their care.

APPENDIX

Because of its importance as Newton's most complete analysis of the universe of stars, there follows a transcription of the draft Proposition (ff275^r, 275^v, 276^r, 184^r, 185^r, 184^v, ULC Add. MS 3965) in its final form:

Prop XV. Theor XV

[*Statement of Proposition.*] Stellae fixae in coelis quiescunt et immensis tam ab invicem quam a Sole nostro intervallis distant.

[*Absence of observed annual parallax shows the stars are at great distances from the Sun.*] Cum nulla sit fixarum parallaxis sensibilis ex Terrae motu annuo oriunda necesse est ut ingens sit earum a systemate nostro distantia. Parallaxis utique minutorum viginti secundorum diametrum circuli quem stella polaris circa polum describit alternis vicibus augeret ac diminueret ac tempore verno majorem redderet quam tempore autumnali minutis octoginta secundis. Et tanta magnitudinis variatio facile sentiretur. Imo variatio quam parallaxis quintuplo vel etiam decuplo minor induceret, ope Telescopiorum sentiri posset. Ex parallaxi autem adhuc minore consequeretur distantias fixarum a Systemate nostro vicibus plus centum millibus majores esse quam distantiam Solis a Terra. Qui minores esse volunt, doceant earum parallaxim sensibilem existere.

[*The stars, being relatively at rest at huge distances from us yet free to move, must be absolutely at rest.*] Fixas autem circa systema nostrum non revolvi triplici constat argumento, tum quod datas servent positiones ad quiescentia Orbium nostrorum Aphelia et Nodos, tum quod nulla extet vis naturae qua corpora tam longinqua a motibus rectilineis perpetuo retrahi et in Orbibus circa nostrum Systema revolventia retineri possint, tum denique quod in coelis fluidis quiescant inter se. Quietis mutuae tantorum tam incomparabilium ac tantis intervallis distantium corporum causa nulla alia in coelis fluidissimis cum ratione excogitari potest quam vera et absoluta omnium quies.

[*The stars, being at rest, must be at similarly great distances from each other. This we confirm by consideration of the numbers of stars of successive magnitudes.*]
Ex eadem fixarum quiete inter se colligitur insuper quod earum corpora non solum [a] Systemate nostro sed etiam a seinvicem ingentibus intervallis [distant. Cum autem corpora] Solis et fixarum in se mutuo pro ratione distantiarum gravia sint, nec tamen cadant in se invicem nec motu aliquo sensibili inter se ex gravitate illa cieantur: necesse est ut corpora fixarum tam a seipsis quam a sole nostro tam ingentibus intervallis distant ut vis gravitatis qua se mutuo petunt plane insensibilis evadat. Ad ingentes autem distantias tam a seinvicem quam a Sole per universa coela spargi sic etiam colligo.

Si corpora lucida innumera & propemodum aequalia ad aequales circiter ab invicem distantias per coela universa spargi intelligantur et ex eorum aliquo ceu centro spectentur caetera: quae huic undique proxima sunt ob minimam distantiam lucidissima apparebunt et erunt quasi duodecim vel quatuordecim numero ut computanti facile patebit; et tot circiter sunt stellae fixae magnitudinis primae, nimirum quindecim. Deinde quae his undique proxima sunt, ob distantias duplo majores apparebunt quadruplo obscuriora et quadruplo plura: et tot circiter sunt stellae magnitudinis secundae, nimirum sexaginta. Quae vero tertio loco circumponuntur ob triplam distantiam erunt noncuplo obscuriora et noncuplo plura: stellae autem magnitudinis tertiae sunt paulo plures nempe 200 circiter: cujus rei ratio esse videtur quod huic ordini adnumerentur quae non solum noncuplo sed etiam decuplo vel undecuplo obscuriores sunt quam stellae magnitudinis primae. Sic enim progressio decrementi lucis singularum magnitudinum fiet magis aequabilis quoad sensum. Quae quarto loco circumponuntur ob quadruplam distantiam erunt sexdecim vicibus obscuriora et totidem vicibus plura: stellae autem magnitudinis quartae ob rationem jam dictam sunt adhuc plures nempe 500 circiter. Sic numerus fixarum magnitudinis cuiusque pro decremento lucis atque adeo pro incremento spatij semper ulterioris quod occupant perpetuo augetur. Nullus enim dubito quin eadem progressio in stellis magnitudinis etiam quintae et sextae alijsque minoribus obtineret si modo illae omnes ab Astronomis notatae essent et in Tabulas relatae.

[*Telescopic stars seem to extend without limit, as required.*] Adhibitis enim Telescopijs tam ingens et prope infinitus numerus minorum et ulteriorum stellarum detegitur: et numerus ille melioribus Telescopijs in tantum semper augetur ut natura in hac progressione limitem minime novisse videatur.

[*We do allow for variation among the stars; the larger stars are separated by greater spaces from their neighbours.*] Fixarum tamen distantias magnitudinibus apparentibus proportionales esse minime statuendum est. Ob diversitatem verarum magnitudinum possunt aliquae magnitudinis primae in coelis earum quae secundae vel tertiae sunt magnitudinis locari et contra. Nam quo majora sunt earum corpora eo magis distare debent ob majorem vim gravitatis qua sese alias attrahent.

[*The Sun is one of the stars.*] Cum fixae omnes non solum nudis oculis sed etiam per optima telescopia, (fumo juxta oculum ad tollendos errantes radios illinita,) instar punctorum mathematicorum appareant, et tamen intensa luce ad distantias tam immensas fulgent: certum est quod luce non reflexa ad modum Planetarum sed nativa ad modum Solis splendent. Si a regionibus nostris in eas fixarum abire liceret, non dubium est quin Sol prout ab eo longius recederemus minor semper appareret donec tandem instar stellae fixae cerneretur: Stella autem omnis fixa ad quam satis accederemus major semper appareret donec Solem nostrum apparenti magnitudine aequare videretur. Igitur Sol et Stellae fixae nativa luce similiter fulgent, et similibus intervallis per coela sparguntur, et in coelis suis similiter quiescunt, et quantum patet per omnia sunt similes, ideoque per Hypoth [IV] pro eodem corporum genere habendi sunt. Unde etiam fixam omnem pro more Solis nostri systemate Planetarum [cingi] a ratione non est alienum.

[*We follow James Gregory in comparing the apparent brightnesses of the Sun and stars and so deriving their relative distances.*] His ita se habentibus, distantiae fixarum ex proportione Lucis praeterpropter innotescunt, computando scilicet distantiam in qua Sol noster instar stellae fixae appareret. Computandi rationem indicavit Jacobus Gregorius at distantiam Gregoriana majorem sic coll[ig]imus. Annulus Saturni cum latissimus apparet tantum lucis reflectit ad nos quantum Saturnus ipse circiter, et uterque simul tantum in quadratura Solis quantum Saturnus reflecteret si foret ejus apparens diameter 26″ circiter. Unde Saturnus una cum annulo excipit $\frac{1}{1000000000}$ partem circiter totius lucis solaris. Ponamus quod Saturnus (Planeta utique plumbei et obscuri coloris) reflectat quintam circiter partem lucis in ipsum incidentis, et lux quae ab ejus hemisphaerio lucido reflectitur, erit $\frac{1}{2500000000}$ pars circiter lucis ab hemisphaerio Solis manantis. Ergo cum lux sit reciproce ut quadratum distantiae corporis lucidi, si Sol 50000 vicibus magis distaret a nobis quam Saturnus, hic aeque lucidus ac Saturnus appareret: hoc est triplo vel quadruplo lucidior ad minimum quam stella primae magnitudinis. Nam stellam primae magnitudinis non minus luce superat quam haec superat stellam secundae. Igitur si duplicetur inventa distantia habebitur distantia fixarum 100000 vicibus major quam distantia Saturni a Sole, id est quasi decies centies mille vicibus major quam distantia Solis a Terra.

[*The method is not undermined by interstellar obscuration.*] Si quis dicat lucem fixarum in transitu per coelos impediri et multum diminui: respondeo quod defectus lucis inde oriundus facile animadverteretur. Nam si verbi gratia tres tantum quartae partes lucis fixarum primae magnitudinis in transitu ad nos perirent, tot partes lucis fixarum secundae magnitudinis ob duplam distantiam bis perirent, ideoque hae stellae transmitterent decimam sextam tantum partem lucis suae ad nos. Et ob eandem rationem fixae magnitudinis tertiae, quartae, quintae et sextae transmitterent tantum $\frac{1}{64}$, $\frac{1}{256}$, $\frac{1}{1024}$, $\frac{1}{4096}$ partem lucis suae ad nos: ideoque cum fixis primae magnitudinis collatae apparerent longe minores quam revera apparent, vel potius non omnino apparerent ob defectum lucis. Nam cum stella sextae magnitudinis ob distantiam sextuplo vel potius octuplo vel noncuplo majorem quam distantia stellae magnitudinis primae, fit quadraginta vicibus obscurior et amplius quam stella illius magn., et insuper ob lucem impeditam et amissam fit 1024 vicibus obscurior: haec erit plusquam 40960 vicibus obscurior.

[*The method is not undermined if the Sun is not an average star.*] Dici etiam potest quod etsi Sol noster sit de genere fixarum tamen ob varias fixarum magnitudines ac distantias ab invicem et ignotas proportiones ad Solem nostrum fieri possit ut earum distantiae sint multo majores vel multo minores quam pro computo superiore: at respondeo quod cum tot fere sint stellae magnitudinis primae quot globi propemodum aequales circa globum ipsis aequalem immediate circumponi possint ac tot secundae quot globi proxime circumstent & sic deinceps: verisimile est distantias fixarum a Sole mediocres esse inter eas quas habent a seipsis aut forte ob numerum aliquanto majorem stellarum quam globorum, magnitudinem mediocrem paulo superare, puta in ratione sex ad quinque; et propterea Solem esse mediocris circiter magnitudinis inter fixas, et mediocri luce splendere. Si lux ejus noncuplo major esset vel noncuplo minor quam pro ratione mediocri, distantia inde collecta non nisi triplo major vel triplo minor evaderet; et ejusmodi minutias hic negligo.

Corol. Igitur ob immensas fixarum' distantias tam a Luce quam a defectu parallaxeos collectas corpora illa nullos edent sensibiles effectus in regione systematis nostri ad perturbandos motus Planetarum.

REFERENCES

1. Voltaire, *Elemens de la philosophie de Neuton* (Amsterdam, 1738), 383.
2. "Continuo opus esse miraculo ne Sol et fixae per gravitatem coeant", David Gregory memorandum no. 44 (Royal Society Library), available in *The correspondence of Isaac Newton* [hereafter: Newton, *Correspondence*], ed. by H. W. Turnbull *et al.* (Cambridge, 1959–), iii, 334.
3. The first evidence of 'proper motions' of stars was given by Halley, "Considerations on the Change of the Latitudes of some of the Principal Fixt Stars", *Philosophical transactions*, xxx (1717–19), 736–8.
4. On the subject of Newton and providence, see I. Bernard Cohen, "Isaac Newton's *Principia*, the Scriptures, and the Divine Providence", in *Philosophy, science, and method*, ed. by Sidney Morgenbesser *et al.* (New York, 1969), 523–48; Henry Guerlac and M. C. Jacob, "Bentley, Newton and Providence", *Journal for the history of ideas*, xxx (1969), 307–18; David Kubrin, "Newton and the Cyclical Cosmos: Providence and the Mechanical Philosophy", *ibid.*, xxviii (1967), 325–46; and Margaret C. Jacob, *The Newtonians and the English Revolution* (Hassocks, Sussex, 1976).
5. The history of *De systemate* is discussed in I. B. Cohen, *Introduction to Newton's 'Principia'* (Cambridge, 1971), 109–15 and 327–35.
6. In the familiar Motte-Cajori translation of the *Principia* (Berkeley & Los Angeles, 1934), 596–7.
7. See Section A, chap. 3.
8. *Principia*, ed. Motte-Cajori, 597: "Tantis igitur intervallis ab invicem distantia sidera nec trahent se mutuo sensibiliter, nec a Sole nostro trahentur" (University Library, Cambridge, Add. MS 3990, f35^r).
9. Newton, *Principia* (London, 1687), 420; ed. Motte-Cajori, 422.
10. Copies of the printed versions of the relevant sermons, and of the 1756 edition of the four letters from Newton to Bentley, are conveniently available with an introduction by Perry Miller in I. B. Cohen (ed.), *Isaac Newton's papers and letters on natural*

philosophy [hereafter: Cohen, *Newton's papers and letters*] (Cambridge, 1958). A more accurate text of the letters from Newton and the one surviving letter from Bentley to Newton are to be found in Newton, *Correspondence*, iii.

11. Bentley's seventh sermon (and the second on "A Confutation of Atheism from the Origin and Frame of the World") (London, 1693), 4; Cohen, *Newton's papers and letters*, 316. *Cf.* Bentley to Newton, 18 February 1692/3, Newton, *Correspondence*, iii, 246.
12. Most notably in the General Scholium in the second (1713) and later editions of the *Principia*.
13. Cohen, *Newton's papers and letters*, 281–2; Newton, *Correspondence*, iii, 234.
14. Cohen, *Newton's papers and letters*, 292–3; Newton, *Correspondence*, iii, 238.
15. Cohen, *Newton's papers and letters*, 293; Newton, *Correspondence*, iii, 239. In the printed version of his seventh sermon, Bentley criticizes the position he originally held: Cohen, *op. cit.*, 351–2.
16. Cohen, *Newton's papers and letters*, 295–6; Newton, *Correspondence*, iii, 239.
17. See for example C. V. L. Charlier, "How an Infinite World may be Built Up", *Arkiv für Matematik, Astronomi och Fysik*, xvi, no. 22 (1922).
18. H. Seeliger, "Ueber das Newton'sche Gravitationsgesetz", *Astronomische Nachrichten*, cxxxvii, no. 3273 (1 February 1895), cols 129–36.
19. Newton, *Correspondence*, iii, 246–53.
20. *Ibid.*, 250–1. The second paragraph, being a comment, is in square brackets in the original.
21. Cohen, *Newton's papers and letters*, 351.
22. The drafts are in the first '*Principia* box', University Library, Cambridge, Add. MS 3965; *all manuscript references are to this box unless otherwise indicated*.
23. f280^v: "Stellae fixae quiescunt et immensis tam ab invicem quam a Sole nostro intervallis distant."
24. ff280^v–280^r: "Datas enim servant positiones ad Aphelia & Nodos ideoque quiescunt: Et nisi quiescerent non manerent in coelis suis. Nam cum nulla sit earum parallaxis sensibilis ex Terrae motu annuo oriunda, necesse est ut ingens sit earum distantia a systemate nostro, et nulla est vis naturae qua corpora tam longinqua in orbibus circa Terram revolventia retineri possi[n]t & a motibus rectilineis retrahi. Abirent igitur in tangentibus orbium." The last two words may be seen in Figure 1. Note that in these transcriptions of Newton's Latin, occasional phrases are lacking because of damage to the original MS but these can usually be supplied with complete certainty from other drafts.

A manuscript intimately related to the sequence we are now considering, namely f175^r, consists of rough drafts of proofs that the stars are very distant, and is notable for its discussion of the 'double star' method of detecting annual parallax that was to be so important in the nineteenth century: "Ingens fixarum distantia concluditur ex defectu parallaxeos orbis magni. Nam Fixae majores atque adeo propinquiores respectu minimarum illarum quae solis Telescopijs videntur nullum habent ex parallaxi motum sensibilem."
25. Which minimum distance he argues (f279^v) is 12,000 times the distance of the Earth from the Sun.
26. J. Gregory, *Geometriae pars universalis* (Padua, 1668), 148. *Cf.* R. de Villamil, *Newton: The man* (London, [1931]), 106.
27. f279^v: "Unde facile colligitur quod Sol distantia ejus a Terra 900000 vel numero rotundo 1000000 vicibus circiter augeretur."
28. ff279^v–279^r.
29. f279^r: ". . . sed ejusmodi minutias negligo."
30. For the original draft, see Figure 1.
31. Kepler's discussion (*Epitome*, Book I, Part II) is quoted by Alexandre Koyré in *From the closed world to the infinite universe* (Baltimore, 1957), chap. 3. Twelve is the maximum number of stars or points satisfying our conditions. *Cf.* Newton, *Correspondence*, iii, 321.
32. f275^r.
33. f276^r.
34. f74^r (see Figure 2), continued on the verso.
35. In his copy of Vincent Wing's *Astronomia Britannica* (London, 1669), now in the Library of Trinity College, Cambridge, Newton has written at the end of the catalogue of fixed stars (p. 263 of the tables):

Magnitudinis primae   16
             secundae 55 iuxta Hal. vel 62 juxta Wing
             tertiae   227
             quartae   506.

Several of the figures have been corrected, suggesting that Newton counted and rechecked the addition himself, and these may be the figures Newton now uses. H. W. Turnbull (Newton, *Correspondence*, ii, 394) draws attention to ULC Add. MS 3958, ff38r–40v, where Newton has compiled "A Table of ye fixed Starrs for ye yeare 1671. Of ye three first Magnitudes", with totals of 13, 43 and 174 respectively. We shall encounter other rough totals in our discussion of "Related Newton documents", below.

36. The parallel between the path trodden in secret by Newton and that followed publicly by Herschel a century later is astonishing. Herschel in 1781 adopted two postulates: "1. Let the stars be supposed, one with another, to be about the size of the Sun. 2. Let the difference of their apparent magnitudes be owing to their different distances, so that a star of the second, third, or fourth magnitude is two, three, or four times as far off as one of the first" ("On the Parallax of the Fixed Stars", *Philosophical transactions*, lxxii (1782), 82–111, pp. 104–5). When challenged on the second postulate, Herschel added a remark that he "rather meant the order into which the stars *ought to be* distinguished than that into which they *are* commonly divided" (*ibid.*, 105); Newton claims his predictions would match the observations "si stellae in magnitudines juxta proportione lucis inversa quadratorum numerorum progressione $1, \frac{1}{4}, \frac{1}{9}, \frac{1}{16}, \frac{1}{25}, \frac{1}{36}, \frac{1}{49}$ designatas distinguerentur" (f74r, our Figure 2). Near the end of his life, when he had hit upon the technique of using two exactly similar telescopes fitted with a range of diaphragms in order to measure the relative brightness of stars, Herschel could at last afford to put at risk the uniform distribution model plus second postulate, by testing it against the numbers of stars of successive magnitudes ("Astronomical Observations, and Experiments Tending to Investigate the Local Arrangement of the Celestial Bodies in Space", *Philosophical transactions*, cvii (1817), 302–31). In the course of an elaborate argument, he considers spheres surrounding the Sun at distances 1, 3, 5, ... and assigns the space between the first two spheres to first-magnitude stars, between the next two to second-magnitude stars, and so on. He then expects numbers of stars of successive magnitudes to be equal to the differences between successive pairs in the sequence 1, 27, 125, ...; that is, to 26, 98, 218, ..., which is a sequence Newton tries on f280r (Figure 1)! *Cf.* my *William Herschel and the construction of the heavens* (London, 1963), chaps 2 and 5.

37. Edmond Halley, "The Number, Order, and Light of the Fix'd Stars", *Philosophical transactions*, xxxi (1720–21), 24–26; the relevant Journal Book of the Royal Society shows that the paper was read on 16 March 1720/1; that is, 1721, and not 1720 as usually stated. On the derivation of the modern definition, see A. Pannekoek, *A history of astronomy* (London, 1961), chap. 40.

38. The counting of numbers of stars of successive magnitudes is important in modern astronomy, for if the counts are lower than expected this may indicate the presence of a dark absorbing cloud in interstellar space.

   Modern astronomy *defines* magnitudes so that a first-magnitude star shall continue to be 100 times brighter than a sixth-magnitude star, in conformity with the approximate values derived from the traditional classification. Intermediate magnitudes are then defined so that the *ratios* of the corresponding figures for successive magnitudes are always the same: that is, $\sqrt[5]{100}$ or 2·512. The *distances* represented by the magnitudes 1, 2, 3, 4, 5, 6, found by taking the square root of the brightnesses, are then approximately 1, 1·59, 2·51, 3·98, 6·31, 10·00, so that the distances are sometimes less and sometimes greater than the corresponding magnitudes, and for the modern magnitude 3 the distance is substantially *less* than 3. However, the mean modern magnitudes for stars *listed by Ptolemy* as of magnitude 2, 3, 4 are 2·21, 3·28 and 4·35 respectively (Knut Lundmark, "The Estimates of Stellar Magnitudes by Ptolemaios, Al Sûfi and Tycho Brahe", *Vierteljahrsschrift der Astronomischen Gesellschaft*, lxi (1926), 230–6, p. 232), so that, *caeteris paribus*, the corresponding distances will be greater and the numbers of stars *larger than would otherwise be the case*. Of course, stars actually vary enormously in luminosity, and a seemingly-bright star may be many times more distant than a faint one, so that it is in fact hazardous to draw conclusions along these lines until the numbers of stars involved are much larger.

39. Edmond Halley, "Of the Infinity of the Sphere of Fix'd Stars", *Philosophical transactions*, xxxi (1720–21), 22–24. This was read on 9 March 1720/1, "the President in the Chair". The summary in the Journal Book shows that considerations of the *physical* nature of light played a greater role in Halley's thinking than the published paper suggests: "The other objection against an infinite number of stars is from the small quantity of light

which they all give whereas were there an infinite number it should seem to be much more. To this Dr Halley replies that light is not divisible in infinitum and consequently when the stars are at very remote distances their light diminishes in a greater proportion than according to the common rule and at last becomes intirely insensible even to the largest telescopes" (Journal Book XII (1720–26) (copy), 94). A full account of the so-called 'paradox' is given in S. L. Jaki, *The paradox of Olbers' Paradox* (New York, 1969), and a shorter discussion in M. A. Hoskin, "Dark Skies and Fixed Stars", *Journal of the British Astronomical Association*, lxxxiii (1973), 254–61.

40. For this and the preceding quotations, see Figure 2.
41. f275^v.
42. ff184^r, 184^v.
43. A fourth manuscript (ULC Add. MS 4005, ff23^r *et seq.*) is headed "The Mechanical Frame of the World" and the second paragraph reads: "2. The Sun is a fixt star & the fixt stars are scattered throughout all the heavens at very great distances from one another & rest in their several regions being great round bodies vehemently hot & lucid & by reason of the great quantity of their matter they are endued with a very strong gravitating power."
44. f542^v.
45. Published Cambridge, 1962. In a modern printed edition the texts have of course a polished appearance alien to the originals.
46. ULC Add. MS 4005, ff21^r–22^r; Hall & Hall, *Unpublished papers of Newton*, 374–6.
47. ULC Add. MS 4005, ff45^r–49^r; Hall & Hall, *Unpublished papers of Newton*, 378–85.
48. Pannekoek, *op. cit.* (ref. 37), 446. For reasons explained above (ref. 38), the modern definition defines the ratio between successive magnitudes as 2·512. "Suppose we count all stars brighter than successive magnitudes, say 10, 11, 12, etc., in a certain area of sky. If the stars were uniformly distributed in space, the numbers should increase for each magnitude step by a factor of about 4·0. The inverse-square law of brightness requires that two groups of stars differing by one magnitude should have average distances proportional to the square root of their relative brightnesses, or $\sqrt{2\cdot512}$. Then the volume of space probed by the counts should increase in proportion to the cube of the distance, or $(\sqrt{2\cdot512})^3 = 3\cdot98$, which, for a transparent, uniformly populated space should equal the ratio of the numbers of counted stars per magnitude" (D. H. Menzel, F. L. Whipple and G. de Vaucouleurs, *Survey of the universe* (Englewood Cliffs, N.J., 1970), 603).
49. On the 'miracle', see ref. 2 above. On James Gregory's method and the model of a uniform distribution of stars, see David Gregory memorandum no. 33 (Royal Society Library, available in Newton, *Correspondence*, iii, 312), item 13 (which includes a numerical slip); there is no known evidence that David had prior knowledge of James's method. On the relevance of Book I, Prop. LXX, see the Notae by David Gregory on the first edition of the *Principia* (Royal Society Library), where against Book III, Prop. XIV, Corol. 2, he writes: "Sed et praeter immensam distantiam earum positio circum circa effectus impedit ex prop: LXX lib: 1" ("But besides the huge distance, their location on all sides hampers the effects, by Book I, Prop. LXX"). This closely parallels Newton's own addition to Corollary 2 in the second edition: "Quinimo fixae in omnes coeli partes aequaliter dispersae contrariis attractionibus vires mutuas destruunt, per prop. LXX lib. 1."

I owe to Miss Christina Eagles references to two other occasions on which Gregory touches on the problem of the stars and gravity. The first is in a memorandum which mentions "Mr Newton's exceptions against my book" (the *Elementa* of 1702), but which is mainly on other topics and concludes: "The fixt Starrs may move inter se by their mutual actions" (f76^r of the David Gregory memoranda in the Royal Society Library). The second is a slip of paper pasted into Gregory's Notae (f47), headed "Ad prop: VII. Lib. III". After ascribing an understanding of gravity to ancient writers (*cf.* ref. 51), the text continues: "Verum si corpora omnia in omnia sint gravia, Quidni Stellae fixae ex gravitate coeant et concurrant? An continuo opus est miraculo ad hunc effectum impediendum? an in immensa quae intercedit inter eas distantia, languescit gravitas? an potius circa diversa centra rotatae planetarum more revolvuntur." Gregory has subsequently added: "Si mundus esset finitus obtineret haec obiectio: existente vero infinito vim nullam obtinet."
50. David Gregory, *Astronomiae physicae et geometricae elementa* (Oxford, 1702), 159–60, transl. from the second English edn (London, 1726), 288–90.
51. Gregory, *Elementa*, 483; transl. from second English edn, 856. As is well known, Newton allowed Gregory to include in his Preface material drafted by Newton in the 1690s in which Newton credits the ancients with profound scientific and philosophical insight.

In one surviving manuscript (f278r) Newton understands Lucretius, *De rerum natura*, 1, 984–91, in the sense of this Gregory passage, in a sentence he afterwards cancels: "... in materiam omnem circum circa positam et per gravitatem mundi infiniti in aequilibrio sustinentur ne se mutuo ruant."

52. Halley, *op. cit.* (ref. 39), 23.
53. Halley, *op. cit.* (ref. 3).
54. However, examination of the manuscript (Royal Society Library) shows that the word "nearly" before "in aequilibrio" is an interpolation into the original draft. This, placed alongside the summary in Journal Book XII (copy), 93, which has him say that in an infinite system a star is at rest "because it has not any tendency to move one way rather than another", suggests that Halley may simply have thought that stars in an infinite system initially at rest will automatically remain at rest.
55. Newton, *Optice* (London, 1706), Qu. 20: "... Et Quidnam est quod impedit, quominus Sol & Stellae fixae in se mutuo irruant?"
56. For Latin text, see ref. 49. A draft on f236v gives a slightly fuller version: "Quinetiam fixae in omnes coeli partes aequaliter dispersae contrarijs attractionibus vires mutuas destruunt. Nam corpus intra superficiem sphaericam constitutum nullam in partem a viribus superficiei attrahitur per Prop LXX Lib 1."
57. f362r: "... tantae autem sint distantiae fixarum et a Sole et a seinvicem ne systemata eorum in se mutuo cadant" (the draft is transcribed in Hall & Hall, *Unpublished papers of Newton*, 355–9); f152v: "... removendo stellas fixas ad commodas distantias ne cadant in seinvicem" (cf. Cohen, *op. cit.* (ref. 4), 531, where however the draft is said to be for Prop. XLI; it is for the very end of Prop. XLII which leads into the General Scholium).
58. "Elegantissima Haecce solis, planetarum & cometarum compages non nisi consilio & dominio entis intelligentis & potentis oriri potuit. ... Et ne fixarum systemata per gravitatem suam in se mutuo cadant, hic eadem [ad] immensam ab invicem distantiam posuerit" (p. 527).
59. Newton, *Optice* (London, 1706), Qu. 23: "Nam cum Cometae moventur in Orbibus valde eccentricis, undique & quoquoversum in omnes coeli partes; utique nullo modo fieri potuit, ut caeco fato tribuendum sit, quod Planetae in orbibus concentricis Motu consimili ferantur eodem omnes; exceptis nimirum irregularitatibus quibusdam vix notatu dignis, quae ex mutuis Cometarum & Planetarum in se invicem actionibus oriri potuerint, quaeque verisimile est fore ut longinquitate temporis majores usque evadunt, donec haec Naturae Compages manum emendatricem tandem sit desideratura." The Latin is more tentative than the English.
60. Newton, *Correspondence*, iii, 235; Cohen, *Newton's papers and letters*, 286–7.
61. H. G. Alexander (ed.), *The Leibniz-Clarke correspondence* (Manchester, 1956), 14.

POSTSCRIPT: HALLEY AND 'OLBERS'S PARADOX'

We have seen that the details of Newton's world picture involving an infinite and near-regular distribution of stars remained in manuscript, but that he discussed his conception with David Gregory in 1694. Gregory in turn gave some hint of the star counts when, in his lectures published in 1702, he compares the number of observed first- and second-magnitude stars with the geometrically derived numbers 13 and 4 × 13. Now in preparing the text of this book Gregory went to Oxford in May 1701 "to talk with Mr Halley about the whole of my Astronomy";[1] and it is possible that this topic came up. At all events, Halley read two short papers in 1721 at successive meetings of the Royal Society with Newton in the Chair, and these contain the first considered treatment of what we now know to have been the Newtonian universe. In particular, Halley discusses the role of gravity in such a universe, and he mentions (and claims to resolve) the 'paradox' that features so strongly in modern cosmologies: that in such a universe one would expect all the sky to be ablaze with the equivalent of sunlight. Halley's papers were published in *Philosophical transactions*[2] and soon republished in the influential abridgement. They were probably known to J.-P. L. de Chéseaux, who in 1744 gave an accurate statement of the 'paradox' and pointed out that a very slight loss of light as it travels through space would be sufficient to avoid the difficulty.[3] They were certainly known to William Herschel, whose own detailed analysis of star counts recapitulates much of Newton's unpublished investigations.[4] And they are cited by Olbers in the study that has led to the 'paradox' being credited to him. From Olbers, the 'paradox' becomes a commonplace in nineteenth-century astronomical literature (though it was never seen as presenting a serious difficulty).

Halley's papers are therefore the channel through which Newton's private enquiries passed unacknowledged into the common possession of astronomers, and their strange amalgam of confusion and insight is deserving of special study.

Halley begins the first paper, "Of the Infinity of the Sphere of Fix'd Stars", by remarking that the universe of stars is taken as being "actually infinite", and that this is confirmed as powerful new telescopes bring still fainter stars into view. Furthermore, "were the whole System finite; it, though never so extended, would still occupy no part of the *infinitum* of Space...; whence the whole would be surrounded on all sides with an infinite *inane*". In addition, a finite system would collapse gravitationally – an interesting proof that Halley had not seen his own recent discovery of three proper motions as a sign that the stars may form a finite system in orbit about a central position, as Wright, Kant, Lambert and William Herschel were all to do. "But if the whole be Infinite, all the parts of it would be nearly [this word is interpolated in the manuscript] *in aequilibrio*, and consequently each fixt Star, being drawn by contrary Powers, would keep its place; or move, till such time, as, from such an *aequilibrium*, it found its resting place...." Or, in the more succinct account of the meeting in the Royal Society Journal Book, a star remains at rest "because it has not any tendency to move one way rather than another".[5]

So far all is well for those favouring an infinite system of stars. But Halley knows of two objections, both "rather of a Metaphysical than Physical Nature". The first revolves around the meaning of the word 'infinite' and need not detain us. The second is the earliest known statement of 'Olbers's Paradox': "Another Argument I have heard urged, that if the number of Fixt Stars were more than finite, the whole superficies of their apparent Sphere would be luminous." The statement is unequivocal, but the justification is either poorly understood or badly expressed: there would be more stars than there are square seconds in the sky.

Halley's solutions to the objection are two-fold, geometrical and physical. The geometrical solution is flawed, and no commentator has made complete sense of its confused language. We can however see where Halley went wrong. Reasoning, like Newton, in terms of spheres concentric on the solar system, he says of the stars on the successive spheres, that "at a greater distance their Disks and Light will be diminish'd in the proportion of Squares, and the Space to contain them will be increased in the same proportion; so that in each Spherical Surface the number of Stars it might contain, will be as the Biquadrate [fourth power] of their distances". For an accomplished mathematician Halley has made a simple error. At (say) twice the distance, the *apparent* surface area or disk of a star is reduced to one-quarter, while the apparent surface area of the heavenly sphere of course remains the same. Or, alternatively, at twice the distance the *actual* size of the star in cross-section of course remains the same, while the actual surface area of the sphere is multiplied by four. Both calculations naturally lead to the same conclusion, that at twice the distance you can fit four times as many stars onto the heavenly sphere. Halley, however, because he reckons with apparent areas in one case and actual areas in the other, ends up with the 'biquadrate' or sixteen times.

He continues: "Put then the distances immensely great, as we are well assured they cannot but be, and from thence by an obvious *calculus,* it will be found, that as the Light of the Fix'd Stars diminishes, the intervals between them decrease in a less proportion, the one being as the Distances, and the other as the Squares thereof, reciprocally." He seems to be saying that the apparent brightness falls off with the square of the distance (correct), and that for large distances the observed angular separation between two neighbouring stars falls off with the distance (correct). How all this helps, no commentator has yet discovered; but at all events, one basic confusion is evidently present in the earlier part of the argument.

The writer in the Royal Society Journal Book may have been similarly baffled, for he passes at once to Halley's second, and physical, solution to the objection:

> The other objection against an infinite number of stars is from the small quantity of light which they all give whereas were there an infinite number it should seem to be much more. To this Dr Halley replies that light is not divisible in infinitum and consequently when the stars are at very remote distances their light diminishes in a greater proportion than according to the common rule and at last becomes intirely insensible even to the largest telescopes.[6]

This formulation, if it be faithful to Halley's intention, adds significantly to the simple claim of the published text:

> ...the more remote Stars, and those far short of the remotest, vanish even in the nicest Telescopes, by reason of their extream minuteness; so that, tho' it were true, that some such stars are in such a place, yet their Beams, aided by any help yet known, are not sufficient to move our Sense; after the same manner as a small Telescopical fixt Star is by no means perceivable to the naked Eye.

In this version Halley seems to say no more than that after a certain distance, stars make no impression on the eye and so are invisible. Commentators have seen it their duty to criticise Halley yet again, on the grounds that he was well aware that a cluster of stars may be visible although the individual stars of the cluster cannot be seen; and that he should have realised that in the uniform distribution likewise, faint stars may have an effect in aggregate even though individually they cannot be seen. But the analogy is not obvious, since in a cluster the stars appear close together in the sky, whereas in the uniform distribution they appear as far apart as their numbers permit.

In the Journal Book, however, we have the further suggestion that because at great distances light is not divisible *in infinitum,* therefore the inverse square law no longer applies beyond a certain stage. There is a further clue at the end of the second paper, where he says of the light of a star at one hundred times the distance of the nearest stars: "This is so small a pulse of Light, that it may well be questioned, whether the Eye, assisted with any artificial help, can be made sensible thereof." Halley, it seems, accepts that for the nearer stars the simple formulation applies; and that at twice, four times, eight times ... the distance, a star sends us one-quarter, one-sixteenth, one-sixtyfourth ... the light. But eventually we reach a stage where the physical nature of light allows us no more to subdivide the emitted light in accordance with "the common rule". Halley does not here tell us what the physical nature of light is; merely that eventually the straightforward geometrical formulation of the problem is no longer appropriate since for physical reasons the light from very distant stars falls off with more than the square of the distance, and that therefore the objection to an infinite universe of stars is unsound.

In his second paper, "Of the Number, Order, and Light of the Fix'd Stars", read the following week, Halley tells us that in the interim "I have attentively examined what might be the consequence of an Hypothesis that the Sun being one of the Fixt Stars, all the rest were as far distant from one another, as they are from us". It seems probable that he had re-read David Gregory's published lectures, an English translation of which had appeared in 1715. Gregory, as we have seen, visited Newton in 1694, and noted:

> To discover how many stars there are of first, second, third etc. magnitude, he considers how many spheres, nearest, second from them, third etc. surround a sphere in space of three dimensions: there will be 13 of first magnitude, 4×13 of second, $9 \times 4 \times 13$ of the third [error for 9×13]. But there are 15 stars of the first magnitude, 4×15 of the second and so on, other inequalities can compensate here.[7]

In his lectures, Gregory enlarges on this when explaining that some astronomers make the different distances of stars the cause of the different apparent magnitudes.

> And this Opinion is very much favoured by the Number of the Fix'd Stars of the first and second Magnitude. For if every Fix'd Star did the Office of a Sun, for a Portion of the Mundane space nearly equal to this that our Sun commands, there will be as many Fix'd Stars of the first Magnitude, as there can be Systems of this sort touching and surrounding ours; that is, as many equal Spheres as can touch an equal one in the middle of them. Now, 'tis certain from Geometry, that thirteen Spheres can touch and surround one in the middle equal to them, (for *Kepler* is wrong in asserting, in B. 1. *Epit*[*ome*] that there may be twelve such, according to the Number of the Angles of an *Icosaedrum,*) and just so many uncontroverted Stars of the first Magnitude are taken Notice of by Observation....
>
> Again, if it be ask'd how many Spheres equal to the former can touch the first Order of Spheres, ... the Number of these will be found to be 52, or $4 \times$

> 13 ... and nearly as many Fix'd Stars of the second Magnitude have been taken Notice of .... Nor will there be any great Difference from Observation in determining the Number of the Fix'd Stars of the other Magnitudes according to this Method. For the farther Consideration of this Matter, it only remains that we shew, that there is the same Order observed among the Fix'd Stars of the first Rank, as there is between the central Bodies of the Spheres touching and surrounding the inmost Sphere (near whose Center the Eye is placed;) and the same between the Fix'd Stars of the second Magnitude, as there is between the central Bodies of the Spheres second in order from these; and so on. And here indeed the Matter does not go on so well, (which made *Kepler* of another Opinion,) as is evident even from hence, that upon the first cast of our Eyes upon the Heavens, some Tracts of the Firmament appear fill'd with innumerable Fix'd Stars, whereas others are found to be almost empty and void of any.
>
> But there is no great Error committed in the Order of the Stars of the first and second Magnitude, as will appear to any one that makes a Comparison: For there are six Stars of the first Magnitude in the Zodiac, three to the North, and four to the South, nearly, as it ought to be according to this Theory.[8]

Gregory, then, believes that the *number* of stars of successive magnitudes matches the predictions from the geometrical model (though he makes no more attempt to verify this than did Newton in the early stages of his own investigation); but the *order* (that is, distribution) of the stars is regular enough for the first two magnitudes but goes badly wrong with the fainter stars.

Halley's paper might well have been written with Gregory's book open before him. Its title features the words 'number' and 'order', and he is quickly into the geometry of the equal packing of spheres, though he – correctly, as modern research has shown – believes that only 12, and not 13, spheres can be packed in the required manner:

> ...I find, that there cannot ... be more than 13 Points in the Surface of a Sphere, as far distant from the Center of it, as they are from one another: and I believe it would be hard to find how to place thirteen Globes of equal magnitude, so as to touch one in the Center: for the twelve Angles of the *Icosaedron* are from one another very little more distant than from its center; that is, the side of the Triangular Base of that Solid, is very little more than the Semidiameter of the circumscribed Sphere, it being to it nearly as 21 to 20; so that it is plain that somewhat more than twelve equal Spheres may be posited about a middle one; but the Spherical Angles or Inclinations of the planes of these Figures being incommensurable with the 360 degrees of the Circle, there will be several interstices left, between some of the Twelve, but not such as to receive in any part the thirteenth Sphere.

Having corrected the geometry, Halley likewise draws the general inference that "it is no very improbable Conjecture, that the number of the Fixt Stars of the first magnitude is so small, because this superior appearance of Light arises from their nearness; those that are less shewing themselves so small by reason of their greater distance". There are in fact, he writes, sixteen stars indisputably of the first magnitude: four in the Zodiac, four to the north, five to the south and visible from England, and three invisible from England (but seen by Halley when he visited St Helena). "But that they exceed the number Thirteen, may easily be accounted for from the different magnitudes that may be in the Stars themselves; and perhaps some of them may be much nearer to one another, than they are to us; this excess of Number being found singly in the Signs of *Gemini* and *Cancer*."

Persevering with multiples of 13, Halley performs the familiar calculations, but with a notable final twist:

> ...at twice the distance from the Sun there may be placed four times as many, or 52; which, with the same allowance, would nearly represent the number of the Stars we find to be of the 2d magnitude: so 9×13, or 117, for those at three times the distance: and at ten times the distance 100×13 or 1300 Stars; which distance may perhaps diminish the light of any of the Stars of the first magnitude to that of the sixth, it being but the hundredth part of what, at their present distance, they appear with.

Halley seems to imply that he has estimated by eye the hundredfold reduction between first and sixth magnitude, which, as we have seen above, has been turned into a definition in modern astronomy.[9] If so, it is a remarkable achievement: it implies a tenfold increase in distance, whereas John Michell suggested about a twentyfive-fold increase in 1767,[10] and William Herschel used a sixfold increase for much of his career.[11] It seems more likely that Halley, like Newton before him in arriving at an eight- or nine-fold increase, was guided by a comparison between the calculated figures and the numbers in the star catalogues.

This done, Halley terminates his paper by reviving the question of the physical properties of light from extremely distant stars, with which we have already dealt.

For all their defects, then, these two short papers served to make public property the analysis of an infinite and near-regular universe of stars developed by Newton in private and unobtrusively sketched in print by Gregory; and, even if the answers offered were unsatisfactory, they raised the questions of the effects of the emitted light and the gravitational attraction in such a universe.

REFERENCES

1. Royal Society MS 247 (the "Gregory volume"), p. 74.
2. See refs 39 and 37 above.
3. On Chéseaux and Olbers, see the work by Jaki cited in ref. 39 above.
4. Herschel cites the second paper in his "On the parallax of the fixed stars", *Philosophical transactions*, lxxii (1782), 82-111, p. 104. On the parallel between Newton and Herschel, see ref. 36 above.
5. Journal Book XII (1720-26) (copy), 93.
6. *Ibid.*, 94.
7. Newton, *Correspondence*, iii, 317.
8. See ref. 50 above.
9. See ref. 38 above.
10. John Michell, "An inquiry into the probable parallax, and magnitude of the fixed stars", *Philosophical transactions*, lvii (1767), 234-64, p. 242, where he estimates a value between the square roots of 400 and 1000.
11. See ref. 36 above.

3. The Cosmology of Thomas Wright of Durham

Thomas Wright of Durham (1711–86) is known to historians of astronomy for the splendidly-illustrated quarto volume which he published in 1750 under the title *An original theory or new hypothesis of the universe*.[1] In this work he explained the appearance of the Milky Way as due to our immersion in what *locally* approximated to a layer of stars, the light of these stars combining to give a milky effect in directions within the layer.

He is also said to have envisaged our star system as disk-shaped and with the Sun near the centre, and for this he has been hailed as the first of the modern astronomers. This, however, is due to a misunderstanding of both the content and the purpose of Wright's thought. In fact, the recent discovery of cosmological manuscripts composed by Wright in early manhood and in retirement enables us to see *An original theory* as one stage in a lifelong struggle to integrate scientific knowledge into a satisfying religious vision of the universe; and a reading[2] of *An original theory* shows that Wright not only did not but, in a real sense, *could* not accept the disk-shaped model of our star system so often ascribed to him.

1. "A Theory of the Universe" (1734)

Wright came from a modest home,[3] and in astronomy was an autodidact who never succeeded in breaking into the circles of the scientific establishment. But once an unsettled youth was behind him he put his energy and varied talents to the service of the aristocracy to such effect that for the rest of his working life he was able to hover contentedly on the fringes of high society, surveying estates, giving science instruction to noble ladies, and occasionally publishing a book[4] or giving a course of public lectures. His interest in astronomy he owed to a childhood teacher, and in his middle 'teens he had been spending so much time on astronomical studies that, in the words of Wright's own journal, "Father, by ill advice, think him mad. Burn all the Books he can get and endevour to prevent study."[5]

The identities of these books we can only guess,[6] but Wright would certainly learn from them that the stars are self-luminous like the Sun and have their own planets and comets. In Whiston's widely-read *Astronomical lectures*[7] he may have encountered the suggestion that the fixed stars appear disordered only because of our position as observers:

> It is very rational to conclude, that some regular Order hath Place also amongst the Fixed Stars. There may be a certain orderly and harmonious Disposition of the Fixed Stars amongst themselves, when they are beheld from some other proper Place, altho' that Order appears not when they are seen from this Earth.[8]

Another book by Whiston especially likely to appeal to Wright's religio-scientific outlook was *Astronomical principles of religion*.[9] Whiston speaks of "the noblest or invisible Parts of the Creation", he describes "this System of the Universe" as "God's great House, or Family, or Kingdom", and he discusses the location of Heaven, Hell and the various spirits.[10] But whereas for Whiston the evil will one day be transferred to a comet, there to be tormented in the sight of the Blessed, we shall find that Wright begins by identifying Hell with the outer darkness, far from the centre of creation where God is especially present.

Wright had certainly given the *Astronomical principles* careful study before composing his mainly-factual *Clavis cœlestis*, published in 1742,[11] and we shall find echoes of it in Wright's last cosmological writings. It may well have been from *Astronomical principles* that Wright first met the problem posed by universal gravitation to those who believed the system of stars to be finite, that if this system is finite the stars will be in danger of collapsing into their common centre:

> ... since withal, the Sun and Fixed Stars do not revolve about one another, or about any common Centre of Gravity ... it follows, that the several Systems, with their several Fixed Stars or Suns, do naturally and constantly, unless a Miraculous Power interposes to hinder it, approach nearer and nearer to the common Center of all their Gravity; and that in a sufficient Number of Years, they will actually meet in the same common Center, to the utter Destruction of the whole Universe.[12]

When Whiston had penned these lines, the stars were of course no longer regarded as visible points on the heavenly sphere, yet their positions in the sky were as fixed and unchanging as ever. In 1718, however, Halley had discovered that three bright stars were no longer in the positions assigned them by the ancients.[13] The number of these stars was almost negligible, the motions detected were extremely small; but for Wright Halley's discovery was enough to show that the stars may after all be revolving about their centre of gravity, prevented from falling into their centre as are the planets from falling into the Sun.[14] For this conception he was to earn the gratitude of Immanuel Kant.

Until recently nothing was known of Wright's first attempt at a satisfactory cosmology beyond his remarks on the opening page of *An original theory*: "The Hypothesis upon which this new Astronomy is founded, and now reduced into a regular System, was the result of my Astronomical Studies full fifteen years ago". From a footnote it is clear that these studies were not limited to scientific considerations: "The first Scheme of this Hypothesis was plann'd in the Year 1734, representing in a Section of the Creation, eighteen Feet long and one broad, several thousand Worlds and Systems, and a great Number of emblematical Figures ...".

This "scheme" does not survive, but a pair of documents related to it were recently identified in Newcastle Public Libraries.[15] The first, headed "The Elements of Existence, or a Theory of the Universe", is a descriptive title to the scheme, which is

> a section of ye Univers ... comprehending first ye Paradise of imortal spirits in there several Degrees of Glory, surounding the Sacred Throne of Omnipotence. Secondly, the Gulfe of Time or Region of Mortality, in which all sensible beings such as ye planetary bodies are imagined to circumvolve in all manner of direction round the Devine Presence, or y^e

Eternale Eye of Providence. Thirdly, the shades of Darkness & Dispare supposd to be The Desolate Regions of ye Damn^d.[16]

The second document is a much longer "explanation" of this "Theory of the Universe", and may well be the text of a public lecture for which the scheme was a vast visual aid. Wright himself has identified these documents with the source of *An original theory*, for in the manuscripts he has added two notes; after the first "Wrote in ye year 1734: the author being then 22 years old", and after the second:

> This Juvenil Performance was the Produce of the authors Imagination before he had Reap'd any advantages either from Reading or Study, but Prov'd afterwards the foundation of his Theory of ye Univers a much more perfect Work.

Wright has apparently forgotten that he had previously devoted himself to studying astronomy even to the appearance of madness; and the lecture fulfils the hope he had expressed in his journal in 1729, "Conseive may find Ideas of ye Deaty and Creation",[17] for it proves to be an attempt to locate the Creator within the creation, and to integrate the natural with the supernatural.

The universe outlined in the lecture incorporates two features which were to characterize all of Wright's cosmological thinking. First, the centre of universe in the moral order, referred to by such titles as "the Sacred Throne", is also the centre of the universe in the physical order, the gravitational centre and source of the laws of nature. Second, and as a consequence of the existence of such a focal point, both morally and physically the rest of creation is arranged symmetrically about this centre.

In the 1734 lecture these two fundamental truths about the Creator and his creation are given only a simple development. The Sacred Throne is the centre of a sphere within which is the Region of Philisity or Heaven; outside this sphere, but within a larger concentric sphere, is the Region of Probation, a spherical shell in which the Sun and myriads of other stars with their attendant planets and comets orbit about the Sacred Throne; while beyond and outside the larger sphere lies the Region of Punishment, the outer darkness of Hell.

Wright's immediate purpose in the lecture is to impress his audience with the long-term advantages of a moral life, and to force his argument home he shows his listeners their present location in the Region of Probation. Indeed, for greater effect he draws in his scheme the visible stars as they appear to us from Earth, rather than representing them in his cross-section of creation as strict logic would demand. And so

> upon all sides the principal stars of the visible creation are exhibited in their natural order as seen from ye Earth by ye naked eye. Those of ye first magnitude nearest to our own system, and the rest proportionable removed according to their respective phenomena. Beyond these are others more remote crownd by a penumbral shaddow such as we call telescopic stars and again without them more, supos^d to be at immense distance & by no means perceptible to ye human eye. At a certain distance from ye Sun equal to a vissual ray of ye smallest visible star is a faint circle of light terminating the utmost extent of ye visible creation, in a finite view from ye Earth....[18]

Wright, we see, has been led almost by accident to an explanation of the Milky Way, the "faint circle of light". His overall theological purpose has encouraged him to discuss the structure of the universe from general principles, without reference to what astronomers know from observational data; the requirements of his two-dimensional visual aid have led him to consider a plane section of creation; and his skills as teacher and preacher have brought home to him the need to indicate in this plane section how observational data may be reconciled with the structure derived from general principles. We on Earth are completely immersed in one small segment of the shell of stars, and observation can tell us nothing of the overall structure of the universe; we find ourselves surrounded by stars and, as pictured in the plane section, the numerous faint stars merge to give the milkiness of the Milky Way.

But, the reader will object, on this argument we ought to see milkiness in every direction: the Milky Way should cover the whole sky. For Wright's cross-section is forced on him only by the limitations of his two-dimensional visual aid; whereas on general principles we know that the creation is symmetrical about the Sacred Throne and therefore no plane section through the Sacred Throne can be essentially different from any other. Yet the Milky Way which we observe in the sky is unique and defines a unique plane; and a satisfactory explanation of it has therefore still to be given.

It is a measure of Wright's preoccupation with theology that he has quite overlooked this objection. Even in 1742 he still describes the stars as "promiscuously distributed throughout the Mundane Space",[19] and promiscuous distribution in all directions is in plain contradiction with the uniqueness of the plane of the Milky Way. Not until 1750 does Wright admit the alternatives, that the stars are "distributed either promiscuously, or in some regular Order",[20] after which he goes on to show that it is the latter alternative which is correct.

2. *"An Original Theory or New Hypothesis of the Universe"* (*1750*)

It must have been a bitter blow to Wright to realize the flaw in his explanation of the Milky Way, and the effort which the repair demanded is reflected in the prominence he gives the Milky Way in *An original theory*. Yet the general problem is as before: first to derive from extra-scientific principles our knowledge of the overall structure of the universe, and then to reconcile this with the local information supplied by observation. There are two notable developments, however, in that Wright is now prepared to allow the existence of other star systems in the universe, each surrounding its own local Divine Centre, and the star system surrounding a single Divine Centre may have more than one component; this has the effect of concentrating attention away from the problem of Hell. As far as astronomical observation goes, the basic difficulty is to reconcile symmetry about our local Divine Centre with the uniqueness of the plane of the Milky Way.

Wright has two solutions to offer, one of which he prefers and rightly so. This preferred solution makes the shell of stars surrounding the Divine Centre exceedingly thin, and the plane of the Milky Way is then the tangent plane to this shell at the point where the solar system finds itself. The shell itself is vast and gently curving, and so when we look along the tangent plane, we see innumerable faint stars whose light combines to give a milky appearance. When we look

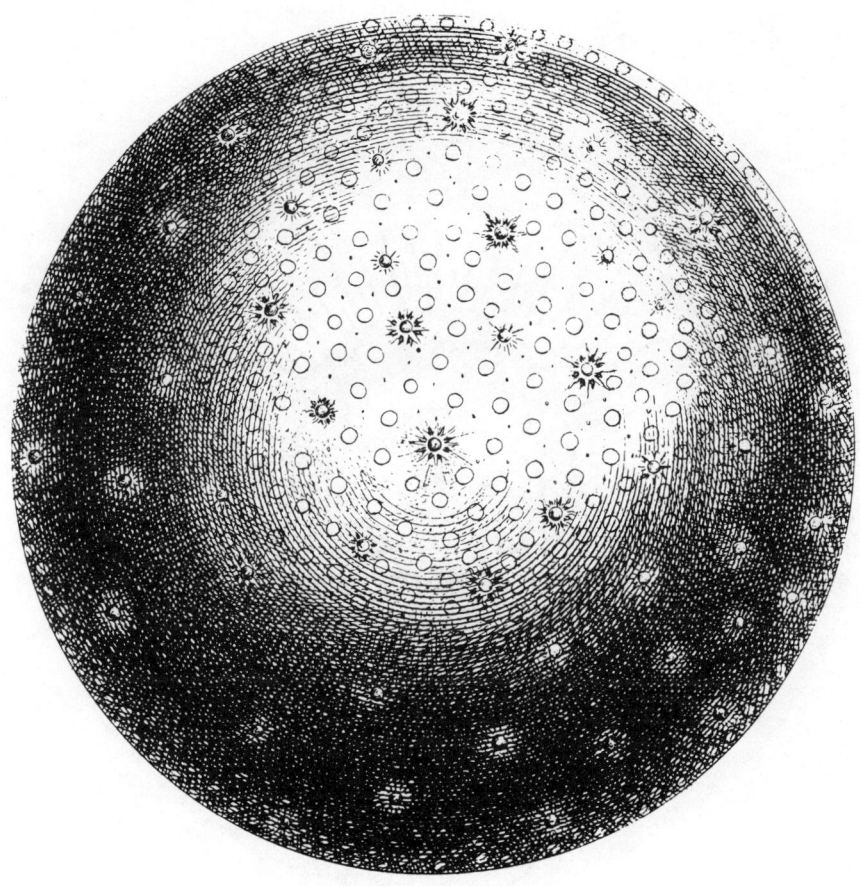

FIG. 1. Plate XXIV of *An original theory* (1750), showing "the intire Creation, as a universal Coalition of all the Stars consphered round one general Center, and as all governed by one and the same Law" (p. 64). This is consistent with the simple picture Wright had held in 1734, whereby the stars (including the Sun) "are imagined to circumvolve in all manner of direction round the Devine Presence, or y^e Eternale Eye of Providence".

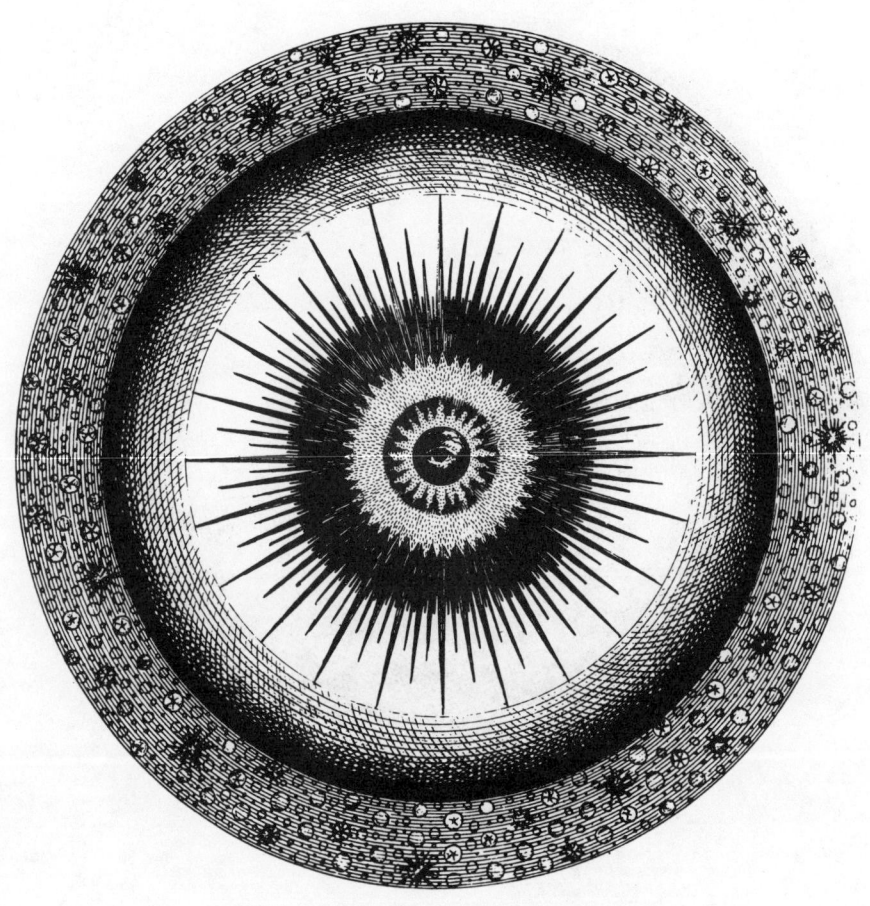

Fig. 2. Plate XXV of *An original theory*, "a centeral Section [cross-section] of the same, with the Eye of Providence seated in the Center" (p. 64).

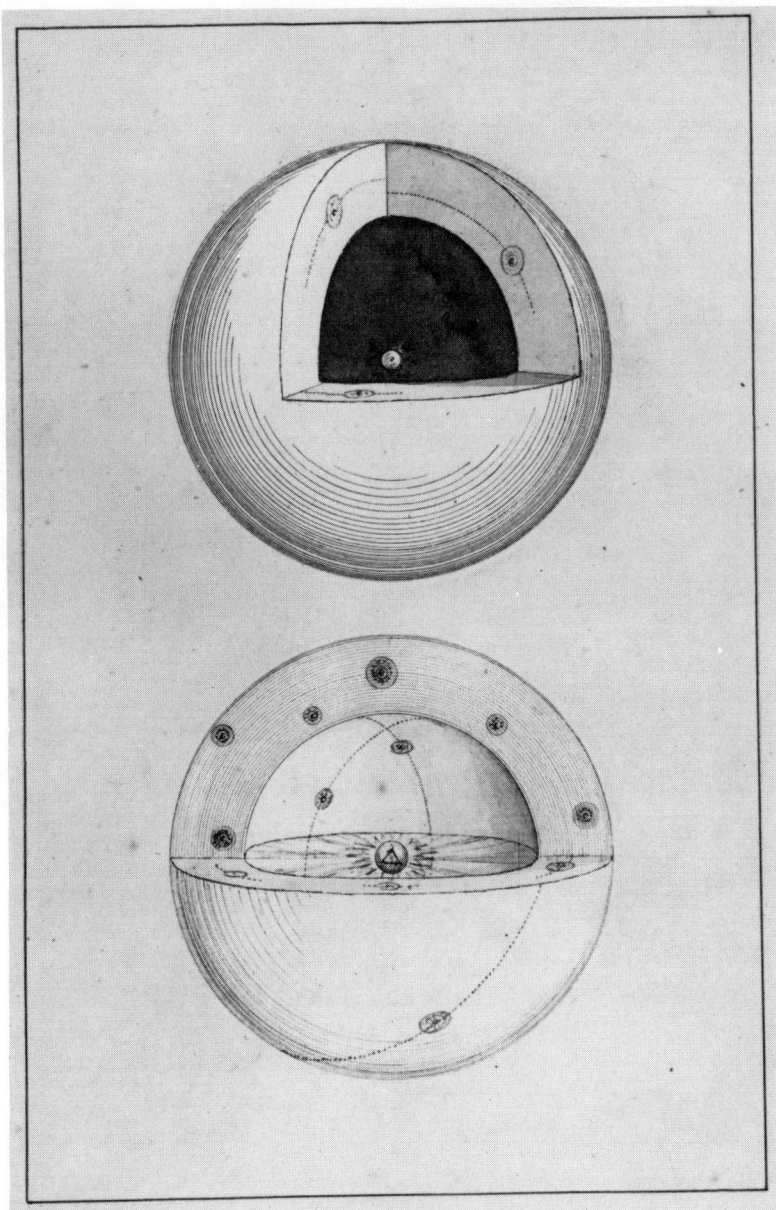

FIG. 3. A sketch by Thomas Wright of the same conception, but showing more clearly the orbits of the stars and their attendant planets around the Sacred Throne (represented by a triangle). (Courtesy of the Curators of Durham University Library.)

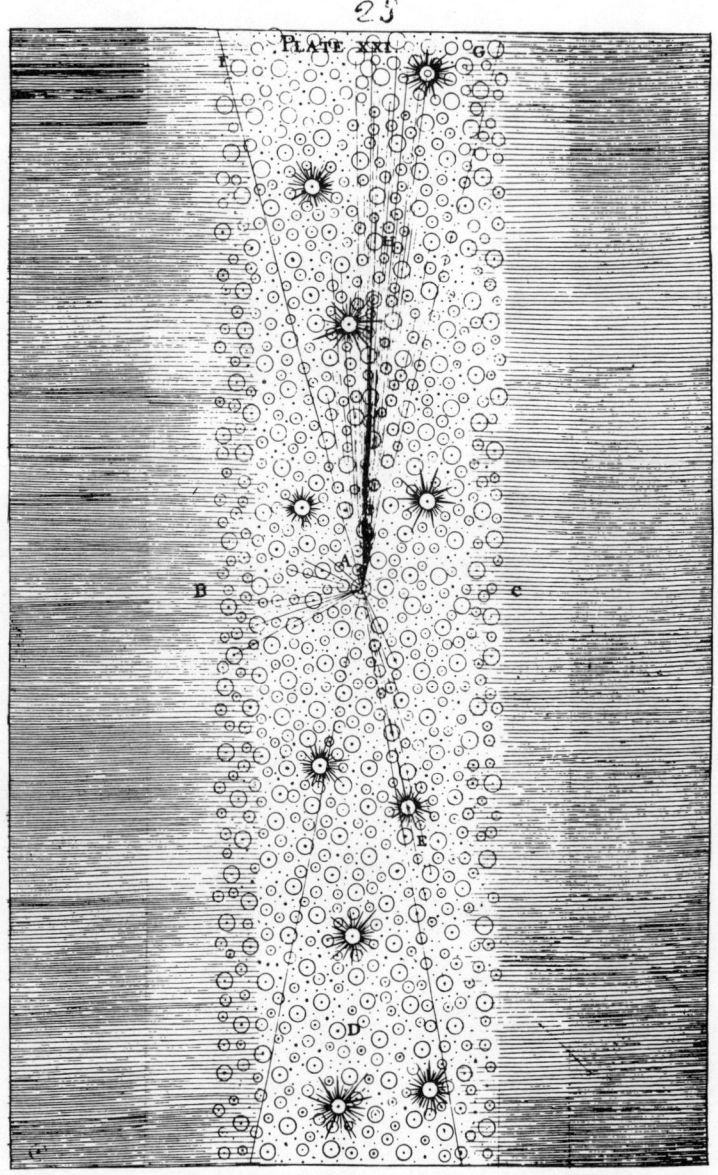

FIG. 4. The much-misunderstood Plate XXIII of *An original theory* (mislabelled XXI). Wright explains to the reader that an observer inside a star system bounded by two parallel planes will observe an effect like the Milky Way. In the *real* universe, the *visible* stars are said to approximate locally to this simplified model. Figures 1-3 may represent the star system to which we belong, provided the spherical shell has a large diameter and a small thickness. The *visible* (nearby) stars then approximate to Figure 4, and the Milky Way defines the tangent plane to the shell in the place where the solar system is currently located as it orbits round the centre of the shell.

Figure I.

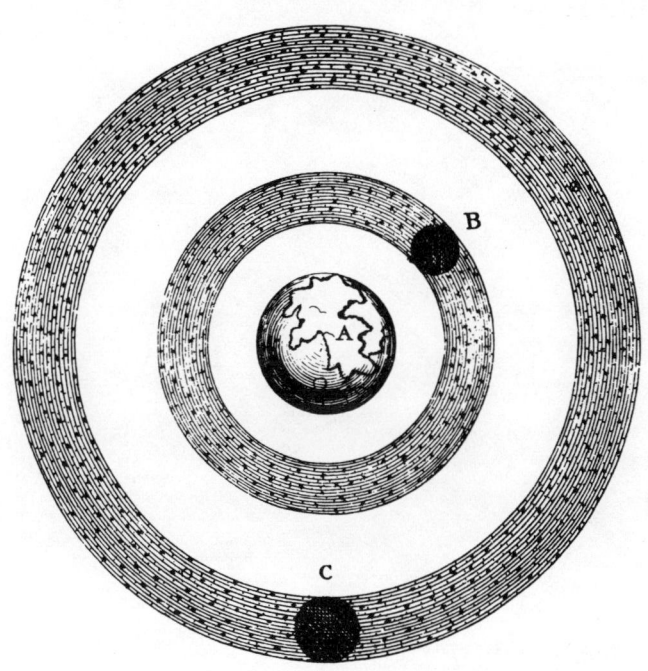

Fig. II.

FIG. 5. Plate XXIX of *An original theory*, representing an alternative explanation of the Milky Way. The stars of our system (both visible and invisible) occupy a space similar to Saturn's ring (or, as represented here, multiple such rings). The *visible* stars form a disk-shaped aggregate, as at *B* or *C*, and so generate the effect of the Milky Way. Wright adds: "... and I cannot help being of Opinion, that could we view *Saturn* thro' a Telescope capable of it, we should find his Rings no other than an infinite Number of lesser Planets, inferior to those we call his Satellites" (p. 65).

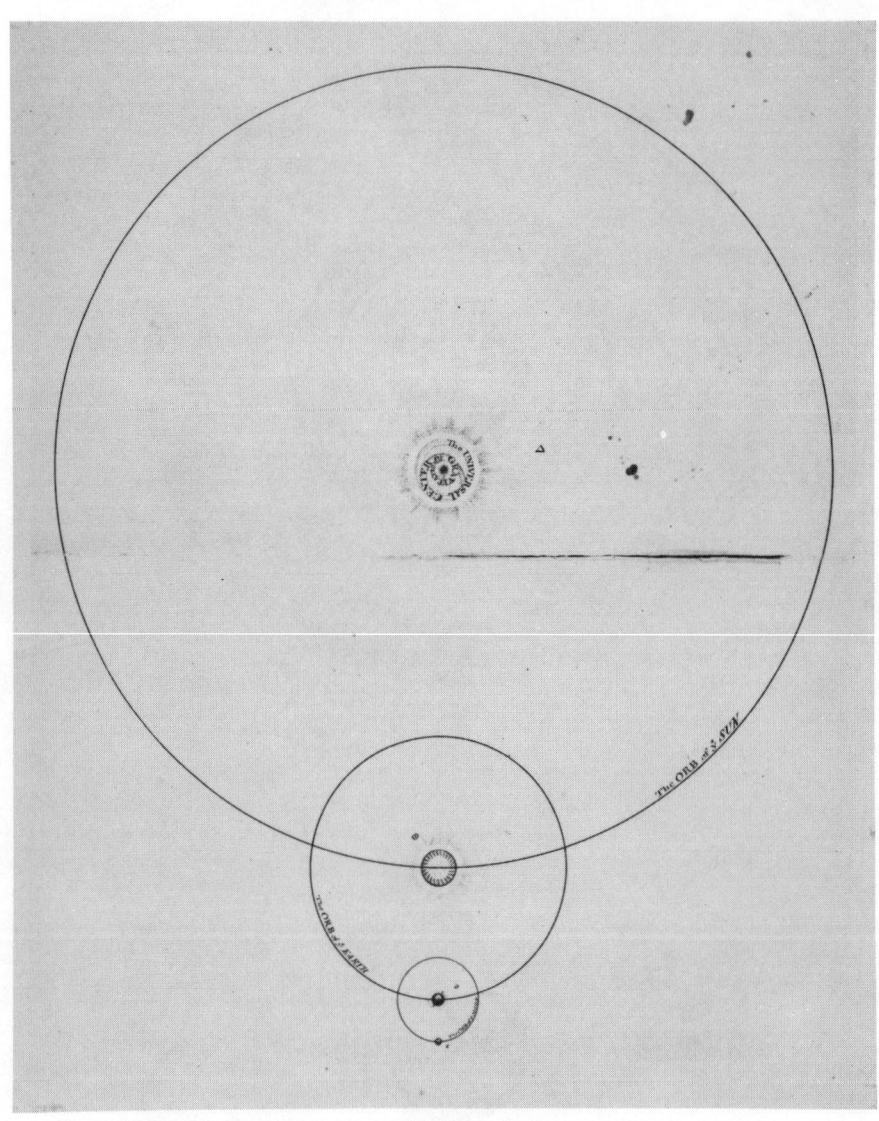

FIG. 6. A sketch by Wright, presumably for Plate XXI of *An original theory,* showing the hierarchy of orbital motions: the Moon about the Earth, the Earth about the Sun, and the Sun about "The Universal Center of Gravitation" (which is also marked with the divine symbol, the triangle). This is perhaps the earliest known drawing showing the Sun in motion. Kant and Lambert were able to extend their hierarchies further because their successive centres were entirely in the natural order. (Courtesy of the Curators of Durham University Library.)

away from the tangent plane, on the other hand, our gaze quickly penetrates through to empty space, and we see only a few stars and those are near and bright. It would follow from this that the Divine Centre lies in the direction of one of the galactic poles.

The other solution has the drawback that it involves the abandonment of full spherical symmetry about the Divine Centre. Instead, all the stars of our system would lie in or close to one particular plane and would form an annulus surrounding the Divine Centre which would also lie in the plane. In Wright's drawings to illustrate this solution, the Divine Centre has the appearance of Saturn while the stars from a flat ring or annulus like the particles which make up Saturn's rings. The *visible* stars, Wright explains, will all lie to one side of the annulus where they form a disk-shaped aggregate; but, on this alternative, beyond the visible stars in one particular direction lies the Divine Centre, and away beyond that are located the stars on the opposite side of the annulus.

In neither solution, we note, does Wright propose anything remotely resembling the disk-shaped star system of modern galactic theory. On the preferred alternative the stars form a thin spherical shell surrounding the Divine Centre; on the other alternative, the stars form a system like a ring or hoop and they again surround the Divine Centre. In either case, the stars circle about the Divine Centre, which is a region in the supernatural order and so devoid of stars. Because of his lifelong commitment to symmetry about the Divine Centre, Wright not only did not, but could not, propose a disk-shaped star-system.

How, then, has the misunderstanding arisen? Does not Wright in the famous Letter the Seventh[21] ask us to imagine that the stars are scattered in the region bounded by two parallel planes, explaining how in this way the appearance of the Milky Way would be produced? Yes, but the word to emphasize is "imagine": Wright, as a good teacher, is showing us the simplest manner in which a milky circle of light might arise, in preparation for the more complex constructions which he is about to propose as genuine possibilities. As he himself immediately says, "But now to apply this Hypothesis to our present Purpose, and reconcile it to our Ideas of a circular Creation".[22]

Unfortunately *An original theory* has been known to astronomers—and historians—mainly through the references in Kant's *Allgemeine Naturgeschichte und Theorie des Himmels*,[23] and Kant himself never saw a copy of Wright's book but was forced to depend upon a summary published in a Hamburg journal.[24] It is not clear from this summary that the Divine Centre is located in the middle of our own star system; Kant, thinking the Divine Centre was supposed remote from our system, considered himself entitled to examine in purely natural terms the proposals of Wright concerning the structure of our system. He knew of the observations by Maupertuis of nebulae which presented an elliptical outline to the observer and which therefore could not be spherical in shape.[25] Kant believed such nebulae to be star systems like our own, and he could therefore discount Wright's arguments that our stars form a spherical aggregate. On the other hand the stars might well, in the words of the Hamburg summary, be "all moving in the same way, and not much deviating from the same plane, as the planets in their heliocentric motion do round the solar body".[26] Little did Kant, in paying tribute to Wright for this conception of the stars as in orbit about a centre, imagine that for him the stars by their motion were preserving the symmetry of our system about its supernatural centre. He therefore saw no reason why the system of stars should not extend without a gap from one edge right across to the other, and he was happy to acknowledge his debt to Wright for the suggestion.

The same misunderstanding enabled Kant to take further steps denied to Wright, and to envisage our own star system as itself but one component of a higher-order system, and so on indefinitely. Wright was acutely conscious of the hierarchy: moon—planet—star—star-system, but at the level of the star-systems, each with its own Divine Centre, the supernatural intervened; the next step would require one Divine Centre to be satellite to another, superior Centre— an impossible proposition; and speculations about systems of systems were therefore denied him.

3. *"Second or Singular Thoughts upon the Theory of the Universe"* (*post 1770*)[27]

From time to time in *An original theory* Wright gives hints of astronomical phenomena—especially novae, variable stars, and nebulae—which were puzzling him as much as they were his contemporaries.[28] In addition, the great earthquake at Lisbon in 1755 set him worrying over the problem of the internal structure of the Earth.[29] We may also surmise that he was less than satisfied with the moral picture he had offered of the universe, with the disordered multiplicity of supernatural centres and the complete lack of any place of retribution for a misspent life.

Of these problems, the explanation of earthquakes presented the least difficulty. Halley had suggested[30] that the crust of the Earth is thin, and that within it a fluid is to be found. This being so, Wright reasoned, the fluid will insinuate itself into the crust, fragments of which will from time to time break off and so set up in the fluid the periodic disturbances which we know as earthquakes.[31] Now Whiston, in his *Astronomical principles of religion*, had speculated about the possibility of creatures living in the interiors of the Sun, planets and comets, and had pointed out that there could be no communication between such creatures and ourselves.

> Nor can they have any Philosophical Evidence that there is such an External World at all; which is the case of the rest of this Universe, as to us, if we, with all the visible Stars, Comets and Planets, be our selves included in such a Cavity; which is not absolutely impossible to be suppos'd.[32]

Indeed not. In fact, the suggestion that we are contained within a solid sky offered an easy solution to Wright's astronomical problems. For if this is so then most of the "stars" are volcanoes pointing towards us so that we may peer into their craters; new and variable stars are volcanoes with changing degrees of activity; comets are formed of material ejected from volcanoes; and the Milky Way

> is looked upon as no other than a vast chain of burning mountains forming a flood of fire surrounding the whole starry regions, and no how different from other luminous spaces but in ye number of stars that compose them, or where there are none, in the vast floods of celestial lava that form it.[33]

These thoughts Wright incorporated in a draft[34] of supplementary material to *An original theory*, possibly with a view to a second edition that never appeared in his lifetime.

In the moral order, however, Wright's conceptions left much still to be desired. The way out of his difficulties was perhaps suggested by the thought that a solid sky which appears set with fiery volcanoes when viewed from within may well have the form of a super-Sun when viewed from without; and that, by analogy, the Sun might be located at the centre of the super-Sun, forming with it two stages in a hierarchy of spheres within spheres.

In developing these ideas, Wright composed several draft essays which he roughly assembled sometime after 1770 under the title *Second or singular thoughts on the theory of the universe*.[35] The manuscript comprises three letters and supplementary material, and is of much the same length as *An original theory*. Fortunately for our knowledge of the development of his ideas, Wright has not edited out his intermediate picture in which the Sun is one of a number of true stars orbiting about the centre of a solid spherical sky set with "stars" of volcanic type, and this is not the only inconsistency: for when his primary concern is with *astronomical* speculation, Wright considers complex possibilities in which each sun or planet may contain within itself an elaborate succession of miniature systems; but when his concern is for the moral order within the universe, he pictures for us a simple hierarchical succession of concentric suns, each with its attendant planets, the succession continuing indefinitely both inwards and outwards.

At long last Wright has achieved his goal of an integration of the moral and scientific universe. The reward of a good life consists in promotion to a more spacious and fulfilled existence in a more ample planetary system; the penalty of a bad life is not crude physical pain but demotion to a more confined life within the Sun. And in the midst of all we find the unique centre in both the physical and the moral order:

> ... the Heaven of one state or creation, may prove little more than the Hades of an other, and so on ad infinitem both ascending to infinity and descending to negation: magnitude and miniature having no proportion or distinction in ye ideas of God, God himself being only the inapproachable center and circumsphere of his own illimitable nature.[36]

Wright's *Second thoughts* do not lack flashes of brilliance: he suggests that undiscovered planets may lie beyond Saturn or inside the orbit of Mercury,[37] he explains the gap between Mars and Jupiter as due to the disintegration of a planet on collision with a comet,[38] and he suggests that the planets may be spiralling slowly into the Sun over millions of years and so may be described as reaching the prime or upon the decline.[39] But to emphasize these passages would be as unhistorical as to blame him for abandoning his explanation of the Milky Way and returning to a solid sky set with stars: in all three cosmological works, Wright's continuing purpose is to integrate the moral and the scientific picture of the universe. Oddly enough, it was his moral purpose that led him to take an outsider's view of our star system and to show that this provided an explanation of visible phenomena including the Milky Way; it was his scientific purpose that led him to explain novae, variable stars, nebulae, comets – and the Milky Way – in terms of volcanoes in the sky.

APPENDIX: WRIGHT AND HERSCHEL

We have seen that although Wright's theological commitments prevented him from taking the final step towards the correct explanation of the Milky Way as the optical effect of our immersion in a continuous layer of stars analogous to the ecliptic plane of the solar system, a reading of a summary of *An original theory* did stimulate Kant to take this step. Is there evidence that William Herschel's 1785 analysis of stellar statistics to establish the shape of the Galaxy (see chap. 1 of Section A) was similarly provoked by a reading of either Wright or Kant?

Herschel nowhere makes acknowledgement to either man, although clearly we cannot exclude the possibility that the idea was passed on to Herschel by letter or in conversation and was remembered by him, consciously or unconsciously. But Herschel owned a copy of Wright's book, which was sold at Sotheby's in 1958 as Lot 452 of the Herschel papers and books and passed into the possession of Mr S. I. Barchas of Tucson, Arizona. Did Herschel acquire this copy *before* assembling material for his 1785 paper on the Milky Way?

On page 19 of *An original theory,* Wright gives the mean distance of the Sun from the Earth as 81 million miles. In the margin of Mr Barchas's copy is written "now proved to be 95 millions". This comment is in William Herschel's hand, but the writing cannot be dated except that it is of his middle rather than old age.[40] Does the figure of 95 millions give us any clue?

In a paper[41] published in 1788, Herschel gives the parallax of the Sun as $8''\cdot63$. This is the value arrived at by Lexell in 1772,[42] and it corresponds to a distance of some $94\frac{3}{4}$ million miles – or in round figures, as Herschel gave it in 1795,[43] 95 million miles.

In a paper read to the Royal Society on 30 January 1781, but unpublished at the time, Herschel states[44] that in April 1779 he observed a sunspot of length 34,476 English miles; and in a paper published in 1795,[45] he tells us that on 19 April 1779 he observed a sunspot $1'\ 8''\cdot06$ in diameter, of length (that is, on the basis of his current value for the mean solar distance) more than 31,000 miles. The sunspots are evidently identical, and a calculation shows that in 1781 he was using for the mean solar distance a value of just under $104\frac{1}{2}$ million miles. This is confirmed by an early notebook[46] where he gives as the distance of the Sun three times 34,761,680 miles, or again just under $104\frac{1}{2}$ million miles. The corresponding solar parallax is under $8''$.[47] In 1782, however, Herschel gives,[48] as a round figure for the angle of the Earth as seen from the Sun (that is, twice the solar parallax), $17''$, and not $16''$ or less. In other words, by 1782 he has abandoned the earlier value, and his new figure no doubt represents (in round numbers) twice Lexell's value of $8''\cdot63$, which corresponds, as we have said, to a distance of nearly 95 million miles.

Herschel's note, then, in his copy of Wright's book could (unfortunately!) have been written at any time from 1781 onwards, either before or after he began his 'gaging' of the Galaxy in the winter of 1783-84. We must therefore fall back on the fact that this is Herschel's *only* marginal note, although he annotated books extensively. In particular, the second time that Wright gives the figure of 81 million miles (page 81) it provokes no comment, suggesting that Herschel was not reading closely at this stage. This would be entirely consistent with his usual attitude to astronomical speculators, and suggests that he glanced through Wright's book when his own studies on the Galaxy were well advanced, expecting to learn little from it, and that his expectations were fulfilled.

REFERENCES

1. *An original theory or new hypothesis of the universe, founded upon the laws of nature, and solving by mathematical principles the general phenomena of the visible creation: and particularly the via lactea*, by Thomas Wright, of Durham (London, 1750). The work aroused very little interest in England, but in Germany there was an abortive attempt to produce a translation (see W. Hastie, *Kant's cosmogony* (Glasgow, 1900), p. lxx). An inferior reprint of the text, but without the fine plates, was issued with additional notes by C. S. Rafinesque (Philadelphia, 1837). A facsimile of the first edition with analytical introduction by the present writer was published in 1971 (London and New York).
2. It is surprising how many readers of *An original theory* have misunderstood Wright. A partial exception is Vera Gushee in her posthumous "Thomas Wright of Durham, Astronomer", *Isis*, xxxiii (1941), 197-218. F. A. Paneth was puzzled as to why his interpretation of Wright's text was inconsistent with the plates (correspondence with Harlow Shapley, in the Harvard Archives).
3. On the first half of his life we are well informed by Wright himself; see Edward Hughes, "The early journal of Thomas Wright of Durham", *Annals of science*, vii (1951), 1-24. Our knowledge of the second half of his life has been transformed by the appearance of Thomas Wright's *Arbours and grottos*, edited with analysis by Eileen Harris (London, 1979).
4. A convenient list of Wright's publications is given in F. A. Paneth, *Chemistry and beyond*, edited by Herbert Dingle and G. R. Martin with the assistance of Eva Paneth (New York, 1964), 114-16. See also the introduction to the reprint of *An original theory*. The Oxford Museum of the History of Science possesses a copy of a late pamphlet by Wright entitled *The longitude discover'd* (Durham, 1773).
5. Hughes, *op. cit.*, 4.
6. A discussion of the likely texts is given in the introduction to the reprint of *An original theory*.
7. William Whiston, *Praelectiones astronomicae* (Cambridge, 1707), translated as *Astronomical lectures* (London, 1715, 1728).
8. Whiston, *Astronomical lectures*, 2nd English edition, 42. Cf. Wright, *An original theory*, 50: "we may reasonably expect, that the *Via Lactea* ... will prove at last the Whole to be together a vast and glorious regular Production of Beings ... and that all its Irregularities are only such as naturally arise from our excentric View"; and *ibid.*, 62: "... in like manner, as the planets would, if viewed from the Sun, there may be one Place in the Universe to which their [the stars'] Order and primary Motions must appear more regular and most beautiful".
9. William Whiston, *Astronomical principles of religion, natural and reveal'd* (London, 1717 and 1725).
10. *Astronomical principles*, 2nd ed., 25, 131-2, 154 seq.
11. *Clavis cœlestis. Being the explication of a diagram entituled a synopsis of the universe: or, the visible world epitomized* (London, 1742). For evidence that Wright had studied Whiston's *Astronomical principles* in writing *Clavis cœlestis*, see the present writer's Preface to the facsimile edition of *Clavis cœlestis* (London, 1967), viii-ix.
12. *Astronomical principles*, 88-89; cf. Whiston, "A discourse ... of the Mosaic history of the Creation", 37, in his *A new theory of the Earth* (London, 1696).
13. E. Halley, "Considerations on the change of the latitudes of some of the principal fixt stars", *Philosophical transactions*, xxx (1717-19), 736-8.
14. In the sixth letter of *An original theory* Wright cites Halley's paper as evidence of motions among the stars and argues that it is by these motions that the stars are prevented "from rushing all together, by the common universal Law of Gravity" (57).
15. By Dr D. M. Knight. The documents are in Volume 7 of the Wright Papers, and a transcription by the present writer is included with the facsimile reprint of *An original theory*.
16. Pp. 2-3.
17. Hughes, *op. cit.*, 5.
18. Pp. C. 2-3.

19. *Clavis cœlestis*, 75.
20. *An original theory*, 30.
21. *Ibid.*, 62.
22. *Ibid.*, 63.
23. This work was ready for publication in 1755, but owing to the bankruptcy of the publisher its release was delayed. The story is told and the work translated into English by Hastie (*op. cit.*, ref. 1).
24. *Freye Urtheile*, Achtes Jahr (Hamburg, 1751), translated by Hastie, *op. cit.*, Appendix B.
25. P. L. M. de Maupertuis, *Discours sur les différentes figures des astres* (Paris, 1742); Hastie, *op. cit.*, 32-3.
26. Hastie, *op. cit.*, 190.
27. Wright gave this title to a carelessly-assembled manuscript which came to light only after the chaotic mass of Wright papers auctioned by Messrs Sotheby on 19 July 1966 were sorted by the present writer. An edition of the manuscript with introduction and notes by the writer was recently published (London, 1968); references below are to this edition. All the papers auctioned in 1966 have since joined the fine collection of Wright material in Durham University Library.
28. *An original theory*, 43. Cf. *Second thoughts*, 21-22.
29. *Second thoughts*, 27.
30. Edmond Halley, "An account of the cause of the change of the variation of the magnetical needle; with an hypothesis of the structure of the internal parts of the Earth", *Philosophical transactions*, xvi, no. 195 (1692), 563-78.
31. *Second thoughts*, 28.
32. Whiston, *Astronomical principles*, 2nd ed., 95-96.
33. *Second thoughts*, 79; cf. *ibid.*, 39.
34. One of numerous drafts incorporated into the manuscript entitled *Second thoughts*.
35. *Cf.* ref. 27. The latest datable item of the manuscript is the reference (1968 edition, 55) to *The nautical almanac for the year 1773* (London, 1771).
36. *Second thoughts*, 63.
37. *Ibid.*, 45, 50.
38. *Ibid.*, 42.
39. *Ibid.*, 72.
40. I am grateful to Mr Barchas for providing photocopies of the relevant pages.
41. W. Herschel, "On the Georgian Planet and its satellites", *Philosophical transactions*, lxxviii (1788), 364-78, p. 368.
42. J. Lexell, "Disquisitio de investiganda parallaxi Solis, ex transitu Veneris per Solem anno 1769", *Novi commentarii*, xvii (1772), 609-72; cited in H. Woolf, *The transits of Venus* (Princeton, 1959), 191.
43. W. Herschel, "On the nature and construction of the Sun and fixed stars", *Philosophical transactions*, lxxxv (1795), 46-72, p. 63.
44. *The scientific papers of Sir William Herschel*, ed. by J. L. E. Dreyer (London, 1912), i, p. cv.
45. Herschel, *op. cit.* (ref. 43), 49.
46. Linda Hall Library notebook, f. 67.
47. Woolf, *op. cit.*, 208.
48. W. Herschel, "On the diameter and magnitude of the Georgium Sidus", *Philosophical transactions*, lxxiii (1783), 4-14, p. 14.

4. *The Cosmology of J. H. Lambert*

Johann Heinrich Lambert (1728-77), the Alsation polymath, is the third of the theological/philosophical speculators of the mid-eighteenth century to achieve an insight into the explanation of the Milky Way. But the three men have met different fates at the hands of history. Kant's *Allgemeine Naturgeschichte* was printed in 1755, and although its distribution was blighted by the bankruptcy of the publisher it is readily available to the modern reader, in German, French, and in recent reprints of the admirable English translation made by W. Hastie at the beginning of the present century. Wright's *An original theory,* as we have seen, was for long known only through the misleading account given of it by Kant, but a modern reprint and editions by the present writer of his other cosmological manuscripts have made him accessible to the historian.

Lambert, however, has remained a shadowy figure. His *Cosmologische Briefe über die Einrightung des Weltbaues* (Augsburg, 1761) is a rare work printed in Gothic type, written in great haste, badly organized, its author "engulfed in cumbersome, long phrases". A French "condensation" appeared in 1770 and 1784 and this was translated into Russian in 1797 and into English in 1800. Most historians have made do with this condensation—or even with the few pages from it included by Milton K. Munitz in his anthology of *Theories of the universe* (Glencoe, Ill., 1957), pages which should surely have been retranslated from the original German of the *Briefe*. Only a few historians have even searched out a copy of the complete French translation (better: paraphrase) by Antoine Darquier, which appeared in 1801 with elaborate notes by J. M. C. van Utenhove.

Happily, the situation is now transformed and the *Cosmological letters,* in translation at least, are accessible to all. First, Alain Brieux of Paris has published a facsimile of the Darquier/Utenhove version, with an elegant and perceptive preface by Jacques Merleau-Ponty in which he rightly emphasizes Lambert's dependence on Leibniz. Utenhove's notes prove to be of considerable interest in their own right, particularly: the long discussion of the nebula in Orion and of the observational evidence in favour of its variability (pp. 129-30, 242); the summary of what little was known with confidence of the proper motions of stars (p. 143); the account of the motion of the solar system through space (pp. 145-7); discussions of the Milky Way (pp. 148, 149) and the nebulae (p. 150); and the account of what could be said of stellar distances (p. 164). These notes offer a perspective on cosmological questions as they were viewed at the very end of the eighteenth century, when Herschel was in mid-career, and they reflect a greater concern for his theorizing than many historians are inclined to allow at this date.

Secondly, Dr Jaki, to whom we are already indebted for whole books on the Milky Way and on Olbers's Paradox, has prepared a meticulous translation into English of the complete work, with extensive editorial apparatus; and it is not too much to say that he has thereby rescued the *Cosmological letters* from the shadows. However, Lambert is an obscure writer at the best of times, and his work was composed in such haste, and in the unhappy format of letters between two friends whose respective roles are not clearly articulated, so that Dr Jaki is often cautious in summarizing the intention of his author. As a result, readers will not find it easy to extract the picture of the universe that Lambert is trying to convey. It therefore seems desirable to attempt such a picture here, although it must be emphasized that we shall confine ourselves to Lambert's hierarchical universe and we shall ignore other, more accessible themes of the *Letters* such as the comets of the solar system.

At the beginning of his Preface to the *Letters*, Lambert refers to the summary of his theory of the distribution of the stars in cosmic space that he had recently given on pp. 505–6 of his *Photometria* (Augsburg, 1760). This admirable and succinct summary, embedded in his discussion of the distances of stars and of Olbers's Paradox, contains the first announcement of Lambert's insights, and deserves to be read even by those unable to invest their time in the *Letters*:

> ... Unde enim *Galaxia*? Mea quidem sententia Systema fixarum, quod se nobis spectandum sistit, haud sphericum sed orbiculare & planum est, atque viam lacteam fixarum veluti Ecclipticam esse pono. Neque enim quendam adfirmatum ire confido, immensum istum stellarum numerum, quem in hoc caeli tractu teloscopiis [*sic*] videmus, ita in isto esse collocatum, ut cunctae stellae quas continet, una sint & minimae, si volumen spectes, & sibi admodum vicinae, si earum a sole nostro distantiam fere aequalem, siue ut rectius loquar, haud infinite diuersam esse ponas.
>
> Porro etsi cunctas istas fixas, quas intueri mortalibus datum est in unum systema complectamur, haud tamen istud erit simplex, sed ex infinitis minoribus compositum. Cunctae istae stellae, quae extra Galaxiam sitae sunt, et maiores, quae in ipso hoc tractu lucent, ad illud systema pertinent, quod solem nostrum comprehendit. Cetera systemata, quae nostro sunt propriora, in ipsa galaxia disseminata sunt. Inaequaliter vero ea disseminata esse vel inde consequitur, quod galaxiae figura admodum est irregularis, atque ista hinc inde dehiscens atque bifariam secta videatur. Solem nostrum non esse in centro sui systematis, inde colligo, quod circulus per mediam galaxiam ductus non est maximus.

By Lambert's standard's this is unusually clear. The visible stars form a system (which for clarity we shall call 'the Galaxy' in what follows) that is flattened and circular, and of this Galaxy the Milky Way is a circle in the sky analogous to the ecliptic that defines the plane containing the planets of the solar system. The Galaxy is made up of endlessly-many smaller systems (which we shall term 'clusters'). One of these contains the Sun together with the stars we see in the sky as remote from the Milky Way and those stars of the Milky Way which are bright (and therefore near). The remaining clusters of the Galaxy are scattered throughout the plane of the Milky Way—irregularly scattered, as we see from the irregular and bifurcated outline of the Milky Way. The Sun is not at the centre of our cluster, for we do not see the Milky Way as a great circle in the sky (and, by implication, the centre of our cluster is exactly in the plane of the Galaxy).

All this makes admirable sense and is carried over into the *Letters* with only one small correction: no doubt reflecting that planets with their systems of moons do move some way from the plane of the ecliptic, Lambert concedes that our cluster may move some way from the plane of the Galaxy, and so he withdraws his proof that the Sun is not at (or near) the centre of our cluster.

What do we know of the genesis of Lambert's conception as expressed in *Photometria*? Later, in 1765, he was to write to Kant that in 1749, "contrary to my habits then, I went into my room after the evening meal, and looked through the window at the stellar sky, and especially at the Milky Way. The insight, which I had then, to see it as an ecliptic of the fixed stars, I wrote down on a quarto page" (*cf.* Jaki, *Letters*, 8). The following year, but unknown to Lambert until after the *Letters* appeared, Wright published his confused explanation of the Milky Way as the optical effect of the light from the *visible* stars of the system to which the Sun belongs, the complete system forming either a thin shell surrounding the Divine Centre or a flattened, hollow ring likewise surrounding the Divine Centre.

As early as 1734, as we know from his manuscripts, Wright had been driven to postulate that the Sun and the stars are moving in orbit around the Divine Centre, so as to avoid gravitational collapse of the system; and at least by 1750 he knew of Halley's 1718 paper announcing that three stars had moved from the positions they occupied in Antiquity. Historians often speak of these movements (in latitude) as the first known proper motions, but to Halley they may well have indicated no more than oscillations about mean positions. Wright was driven by his cosmology to *expect* stable orbits, and interpreted Halley's paper in this sense. Kant had the advantages of sharing Wright's thoughts, and of knowing Bradley's 1748 paper on nutation with its remarks on proper motions and the length of time needed to detect such motions because of the remoteness of the stars. But Lambert seems to have been so struck by the analogy between the ecliptic and the Milky Way that he postulated that the stars are in orbit *even though he himself knew of no evidence of proper motions*. Only after the first version of the *Letters* was completed did he learn of Tobias Mayer's very important announcement in 1760 of numerous proper motions! In this he doubly out-thought Newton, first by appreciating the significance of the Milky Way, which Newton neglected, and second by realizing that the enormous distances of even the nearest stars (confirmed by Lambert in *Photometria*) meant that the lack of known proper motions proved nothing; Newton, although the first to appreciate the distances of the stars from the solar system (strangely, Jaki credits this to Herschel (p. 34)), had failed to make this inference and so had agreed with Richard Bentley that the stars are evidently at rest.

To turn now to Lambert's elaboration in the *Letters* of the conception outlined in *Photometria*, a task that was in hand in June 1760 and completed by the following January. In the tradition of many early-eighteenth-century Newtonians and Leibnizians, Lambert comments on the order to be found in the solar system and lays down the principle (Jaki, p. 106) that "the disorder in the world is only apparent, and where it appears to be the greatest, there the true order is even more excellent though only more hidden to us". Unlike Kant, with whom he has so often been wrongly bracketed, Lambert presents us with a stable universe, in contrast to Kant's vision of gravity working in time to bring structure out of chaos. Like Wright and Kant, Lambert reflects on the hierarchy in nature whereby moons orbit around planets, planets orbit around the Sun and similarly, no doubt, for the other stars, and the Sun likewise is in orbit as part of a larger system of stars. But here the three men part company. For Wright, the centre of the Sun's orbit is in the supernatural order, and although in his 1750 book he allows many such star systems each with its own supernatural centre, the hierarchy inevitably comes to an end. For Kant, the next step in the hierarchy is the Galaxy itself, and from there the hierarchy continues upward *infinitely*; at the centre at every stage is a massive, *luminous* body, and in the Galaxy the bright star Sirius may be the central body. For Lambert, the next step is the star cluster and the step after that is the Galaxy, and although the hierarchy continues upwards for perhaps a thousand steps in all, it is *finite*; further, because the stars are sufficient to give the illumination that is necessary, the massive bodies or 'regents' that are probably but not inevitably at the centre of each system will be *dark*.

In the *Letters*, Lambert adds detail to this conception. Although, as we mentioned earlier, he no longer retains the *Photometria* proof that the Sun is not at the centre of our cluster, the Sun is clearly not at the outside, for then we would see bright stars (which also belong to our cluster) only to one-half of the sky (Jaki, p. 113). Assuming the Sun is some way from the centre, then the centre may well lie towards Orion or Canis Major so that the numerous stars we see in that region of the sky will mostly lie the other side of the centre (p. 131). Indeed the variable light [M42] observed in Orion may be the dark body at the centre illuminated by orbiting stars (p. 166).

Our cluster contains millions of stars (p. 112) as do other clusters (p. 159), and its diameter could be 150 times the distance of Sirius (p. 136). From the appearance of the Milky Way we can infer that our cluster lies "not only somewhat outside the plane of the Milky Way but also closer to its periphery than to its centre" (p. 125). As to the space separating our cluster from its nearest neighbours, "if I put, according to my last letter, the nearest [cluster] in the Milky Way still 10 times farther [than the diameter of our cluster], then I come to 1500 such [Sirius-distances] which consequently would do for about the inner diameter of the Milky Way [*i.e.* Galaxy]" (p. 136). It seems evident from the context that "inner diameter" (*innern Diameter*) refers to the empty space immediately surrounding our cluster, as there is around any ordinary cluster making up the Galaxy. Seen from our point of view, the nearest clusters other than our own encircle us and this "circle" has a size.

Since the clusters are taken as fairly uniform in size and spacing and are all in or close to the plane of the Galaxy, considerations of geometry show that at most six can be immediately around us (p. 124). As there are substantial gaps between one cluster and the next, there must be very many such clusters in all since the Milky Way appears unbroken (p. 125). Indeed, the clusters may be hundreds deep (p. 137), implying (if we convert Lambert's various figures into the equivalent modern units) that the Galaxy has a diameter of millions of parsecs. On the other hand, if the number of clusters were very great indeed, then the Milky Way would seem to divide the sky almost exactly in half as it stretched away to virtually infinite distances, whereas in fact the Milky Way is not a great circle (p. 125). For the same reason our cluster may lie outside the plane of the Galaxy [at the present time], but not far outside, for otherwise the Milky Way would appear more broad than it does (p. 113). The regent at the centre of the Galaxy may, like the regent of our cluster, lie away in the direction of Orion (pp. 138-9). Doubtless the Galaxy belongs with other galaxies to a system of next higher order controlled by an appropriate regent, and so on, "until we finally come to that body which rules the whole creation as its own realm" (p. 168).

Thanks to the efforts of Dr Jaki and Professor Merleau-Ponty and to the initiative of their publishers, we are now able to see Lambert as the creator of an essentially simple and elegant theory of a stable, hierarchical universe; but it was a theory dominated not only by the physics of Newton but also by the philosophy of Leibniz. It applied to cosmology an attitude of mind, a quest for order and stability, that belonged, not to the previous generation, but to the generation before that.

APPENDIX: LAMBERT AND HERSCHEL

Did William Herschel derive elements of his own cosmology from a study of Lambert's *Cosmologische Briefe* (Augsburg, 1761)? It is well known that by 1805, when he referred to the French version of the *Briefe* under its title *Système du monde* and dismissed it as "full of the most fantastic imaginations" ("On the Direction and Velocity of the Motion of the Sun, and Solar System", *Philosophical transactions*, xcv (1805), 233–56, p. 235), he held Lambert's cosmology in contempt. But this does not tell us when he first read the work, nor whether his original reaction was as negative and hostile.

These questions can now be answered from documents in the Herschel Archives of the Royal Astronomical Society. Herschel's Letter-book (MS W 1/1) contains a copy of a letter to his friend Professor Patrick Wilson of Glasgow dated 10 February 1799, from which it emerges that Wilson had a friend who was to translate the *Système du monde* into English, and that Wilson had sent a copy of the book to Herschel for comment. By 10 February Herschel had read it twice, and was part-way through it a third time. The passages he regarded as "exceptionable" he had marked.

> The result of this seems to be that so many places are marked that a republication of a work so liable to *criticism* cannot be of any service to advance the true knowledge of astronomy. The perusal of the Book has put me in mind that my papers [of the 1780s] on the construction of the heavens may now receive a considerable addition by the continued experience I have obtained from observations. Another paper therefore, on that subject is now in my thoughts, and you may easily suppose that many of Mr Lambert's Ideas will be taken notice of, and put in their proper light (p. 236).

In a manuscript book entitled "Miscellaneous Remarks on Books &c" (MS W 7/2), we find more than ten foolscap pages (ff. 17v–22v) of remarks on *Système du monde*, most of them scathingly critical. For example, he comments on p. 140: "The auther [sic] now uses all the license of the poets in the flights of fancy. He confesses that it makes his head giddy, and that he does not know where to stop. I do not call this Astronomy, but wild imagination" (f. 21r). Sometimes he contradicts Lambert from his own observational experience ("106. The author pronounces without the least hesitation or exception that every star has its system of millions of comets, planets, rings satellites &c. Had he seen one of my compressed clusters of stars he would probably have hesitated a little" (f. 19r)), but more often he is contemptuous of his mode of argument ("118. The author is rather at a loss how to place the milky way with respect to the other milky ways, and his reason is not a bad one. For he says he knows of no other milky way" (f. 20r)). There is nowhere a suggestion that Herschel learned anything from Lambert, but in one comment Herschel does manage to encapsulate something of the contrast between the stable, unchanging universe of Lambert and his evolutionary cosmogony: "19. The author hastily concludes from his foregoing arguments against the planets acquiring moons that the comets, the planets and their satellites have *always* moved as they now move, and have *always* been what they are now. This is saying a thing has *always* been what it is because I cannot account for how it *came to be* so" (f. 17ᵛ).

Herschel understands Lambert to argue on p. 120 that "The Sun at the distance of Saturn or still farther from us would be as bright as it is in its present situation", but "This I cannot allow" (f. 20ʳ; *cf.* Lambert's discussion in Letter XIV of the *Briefe*). This evidently provoked him into attempting a definitive solution of such questions in his very next paper, dated 20 June 1799, "On the Power of Penetrating into Space by Telescopes; with a Comparative Determination of the Extent of that Power in Natural Vision, and in Telescopes of Various Sizes and Constructions; Illustrated by Select Observations" (*Philosophical transactions*, xc (1800), 49-85). In the text as published (p. 55), Herschel says: "Optical writers have proved, that an object is equally bright at all distances. It may, therefore, be maintained against me ..."; but in the text as submitted (MS W 5/3.1, p. 8) we find instead "Mr Lambert says, in his *Système du monde*, that an object is equally bright at all distances; and that the sun at the distance of Saturn, or still farther from us, would be as bright as it is in its present situation [and he cites p. 120 of the *Système*]. We cannot be surprised if from a principle so extraordinary he is led to the most unaccountable conclusions ...", and he goes on to develop his critique at length.

Evidently there were protests at the Royal Society at this attack on Lambert, for we find an undated draft letter (MS W 5/3.3):

> Sir, I have the honour of your letter, and can have no objection to the omission of any passages in my paper that you or those of my friends who have been consulted upon the subject may think had better not be inserted. . . . I flatter myself that the inclosed alterations will fully meet with your approbation. . . . My critical observations on Mr Lambert were not occasioned by any disposition of finding fault, but were merely introduced because I thought that a doctrine such as mine which is supported by observations, and which flatly contradicts what has been asserted by a mathemtician of considerable eminence, obliged me to take notice of him. . . . As Mr —, from what he has said to me at the R.S. is probably the Gentleman who objects to my opposing Lambert, I call on him to defend that author's doctrine if he can. . . .

And with this draft is a copy of the amendments which served to remove all mention of Lambert from the printed text.

Years before, when Herschel published his first (1783) paper on solar motion, J. H. Schroeter had written to him and mentioned Lambert's interest in the problem (MS W 1/13.S.15). In the *Système* Herschel now learned with contempt of Lambert's belief that the Sun is in orbit about a massive central body of our star cluster, possibly the Orion nebula; and three years later, in the "Remarks on the Construction of the Heavens" preceding his third catalogue of nebulae and clusters (*Philosophical transactions*, xcii (1802), 477-528), we find what is evidently another critical reference to Lambert, when Herschel says: "... though we may have good reason to believe that our system is not perfectly at rest, yet the causes of its proper motion are more probably to be ascribed to some perturbations arising from the proper motion of neighbouring stars or systems, than to be placed to the account of a periodical revolution *round some imaginary distant centre*" (p. 486; emphasis supplied). After a further three years, Herschel was to return once more to Lambert's suggestion that the Sun is in orbit around the Orion nebula, and this time the Royal Society was to allow Herschel to speak of Lambert's "fantastic imaginations", his first and only criticism of Lambert to appear in print.

To sum up: Lambert's cosmology, as portrayed in the *Système du monde*, was closely and critically studied by Herschel early in 1799. There is no reason to suppose he had anything to learn from it; but his reading revived his own cosmological interests, and the *Système* is reflected in three of the papers he published in the next few years.

BIBLIOGRAPHY

J. H. Lambert, *Cosmological letters on the arrangement of the world-edifice*. Translated with introduction and notes by Stanley L. Jaki (Edinburgh and New York, 1976).

J. H. Lambert, *Lettres cosmologiques sur l'organisation de l'univers*. Reprint of the 1801 translation by A. Darquier, with preface by Jacques Merleau-Ponty (Paris, 1977).

C. THE RIDDLE OF THE NEBULAE

1. *William Herschel's Early Investigations of Nebulae*

The term 'nebula', Latin for mist, cloud, has been used by astronomers since Antiquity to characterise objects similar to the stars except that they present a blurred appearance, unlike the points of light of true stars.

In reality, as we now know, nebulae are of several different types and lie at very different distances. Some are vast, isolated stellar systems, spiral or elliptical, akin to our own Galaxy and separated from us by very great distances – 'island universes' in the nineteenth-century phrase (see chap. 3 below), or 'galaxies'. Then there are star clusters in our own Galaxy, comprising the galactic (or open) clusters, which are irregular associations of stars lying in the central plane of the Galaxy and so usually appearing close to the Milky Way, and the highly condensed globular clusters, which are spread over a large spherical region surrounding the Galaxy. Finally there are the gaseous or galactic nebulae, "true nebulosity" in Herschel's language; these are mostly irregular patches of interstellar gas in the plane of the Galaxy, but a few are remnants of the catastrophic explosion of stars, and there are also the imperfectly understood planetary nebulae (discovered and named by William Herschel) consisting of a sphere or shell surrounding a central star (which may not be readily visible).

Edmond Halley described in *Philosophical transactions* (xxix (1714-16), 390-2) six "luminous Spots or Patches, which discover themselves only by the Telescope, and appear to the naked Eye like small Fixt Stars; but in reality are nothing else but the Light coming from an extraordinary great space in the Ether; through which a lucid *Medium* is diffused, that shines with its own proper Lustre.... In all these so vast Spaces it should seem that there is a perpetual uninterrupted Day...". His six nebulae were the Andromeda Nebula (a galaxy, observed in 1612 but not seen again until 1665; see Section A, chap. 2), the Orion Nebula (gaseous), three globular star clusters, and one galactic star cluster; and we can anticipate great problems for the astronomers we shall study as they try to contain such disparate objects within the confines of a simple theory.

Most astronomers, however, regarded nebulae as star clusters disguised by distance, and it is usually said that this was the view at first taken by Herschel, who was sent a catalogue of sixty-eight nebulae and star clusters late in 1781 and who possessed in his (first) 20ft reflector a powerful tool for the investigation of nebulae. And it must be admitted that in his first great paper on "the construction of the heavens" (1784) he implies that on receiving the catalogue he rushed to examine the nebulae and found that most were clearly revealed as star clusters:

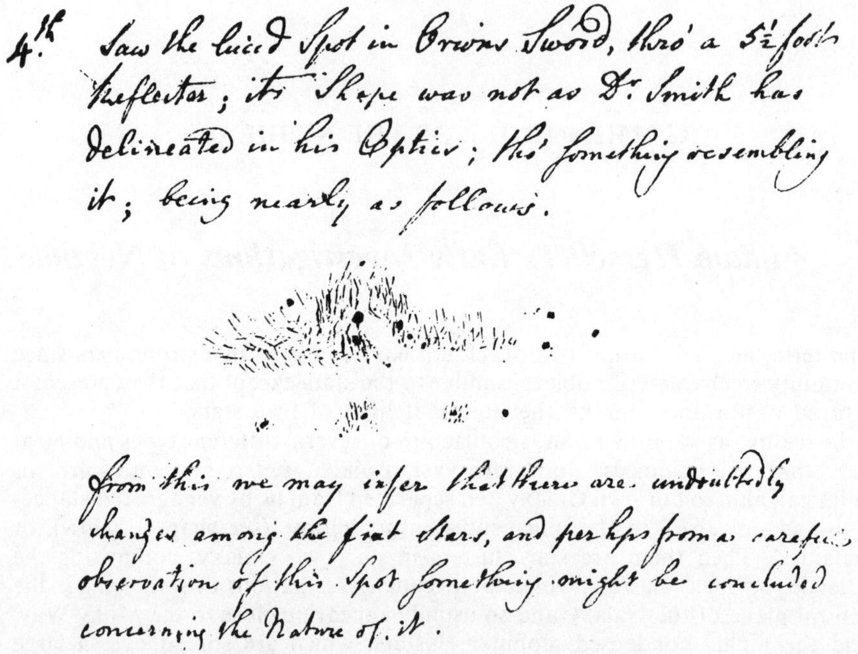

> 4ᵗʰ Saw the lucid Spot in Orion's Sword, thro' a 5½ foot Reflector; its Shape was not as Dr. Smith has delineated in his Optics; tho' something resembling it; being nearly as follows.
>
> From this we may infer that there are undoubtedly changes among the fixt stars, and perhaps from a careful observation of this Spot something might be concluded concerning the Nature of it.

FIG. 1. Herschel's sketch of the Orion Nebula, M42, made on 4 March 1774 (Royal Astronomical Society Herschel MSS, W.2/1.1, f. 1).

As soon as [Charles Messier's 1780 catalogue of nebulae and star clusters] came to my hands, I applied my former 20-feet reflector of 12 inches aperture to them; and saw, with the greatest pleasure, that most of the nebulae, which I had an opportunity of examining in proper situations, yielded to the force of my light and power, and were resolved into stars. For instance, the 2d, 5, 9, 10, 12, 13, 14, 15, 16, 19, 22, 24, 28, 30, 31, 37, 51, 52, 53, 55, 56, 62, 65, 67, 71, 72, 74, 92, all of which are said to be nebulae without stars, have either plainly appeared to be nothing but stars, or at least to contain stars, and to shew every other indication of consisting of them entirely.[1]

The obvious meaning of this is that the nebulae listed were among those Herschel examined as soon as he received the catalogue from William Watson Jr in December 1781.[2] True, the last four nebulae were not in this, Messier's 1780 catalogue, and a manuscript of Herschel's paper[3] shows that nos. 51, 52 and 74 were inserted during the drafting of the paper in the spring of 1784; but this scarcely diminishes the force (and plausibility) of Herschel's categorical statement that he examined the conveniently-situated Messier nebulae as soon as he learned of their existence, and that he satisfied himself that most of those examined were simply star clusters disguised by distance. This after all was no more than to be expected of a man who by immense physical labour and ingenuity had built himself the most powerful instrument in existence for the examination of faint objects such as nebulae;[4] and what more natural than that he should generalize and equate nebulae with star clusters, as he does unequivocally in his "construction of the heavens" papers of 1785 and 1789.[5]

Recently, Herschel's own observing journals became available when the archival collections held by the Royal Astronomical Society were ordered and catalogued.[6] They show that, far from making frequent observations of nebulae in the weeks following the receipt of the catalogue sent by Watson in December 1781, Herschel never looked at another nebula until after his removal in August 1782 from Bath to the neighbourhood of Windsor Castle, where he was to be a professional astronomer available to the Royal Family when required. True, he had every excuse for this. He was in the greatest demand as a musician and conductor of oratorios and as a teacher of music, and on this his livelihood depended.[7] In addition, his election as a Fellow of the Royal Society on 6 December 1781 coincided with the reading of his paper "On the parallax of the fixed stars".[8] In this paper he claimed to be using eyepieces with magnifications not only of hundreds but of thousands, claims which were in fact justified[9] but which provoked outright incredulity and bitter hostility.[10] This and other controversies reduced still further the time available for observations. Soon thereafter he was forced to spend late May, the whole of June and part of July in and around London while his friends worked to secure him the royal patronage which would enable him to devote his life to astronomy. But on the other hand, he chose in late December 1781 to begin another and more extensive 'review' of the stars instead of examining nebulae from the catalogue he had received a few days before; and it is easy to exaggerate the difficulties in his way —in the four weeks beginning 6 February 1782, he observed stars, but not nebulae, on eight evenings.[11] Even when eventually he became a professional astronomer, it was to be several months before he examined more than the occasional nebula. The account of his that we have just quoted is, therefore, very misleading, and we may ask, What *is* the story of Herschel's early encounters with nebulae, and by what steps *did* he come to equate nebulae with star clusters?

Herschel opened his first observing Journal on 1 March 1774.[12] That night he examined Saturn's ring together with the Orion Nebula (M42); he would have known of the Orion Nebula from his copy of Robert Smith's *Opticks* (1738), for Smith mentions the six "lucid spots" orginally listed by Edmond Halley in *Philosophical transactions* for 1715 and cites in particular the Orion Nebula.[13] The entry in Herschel's *Journal* reads:

> Observed the Lucid Spot in Orions Sword belt; but the air not being very clear it appeared not distinct.[14]

On 4 March his entry for the night was nothing short of prophetic:

> Saw the Lucid Spot in Orions Sword, thro' a 5½ foot reflector; its Shape was not as Dr Smith has delineated it in his Optics; tho' something resembling it; being nearly as follows [see Figure 1]. From this we may infer that there are undoubtedly changes among the fixt stars, and perhaps from a careful observation of this Spot something might be concluded concerning the Nature of it.[15]

Smith had reproduced a sketch by Huygens originating from 1656, and it is notable that Herschel seizes at once upon the implications of observations of changes in nebulae—for the history of the study of nebulae, and especially the

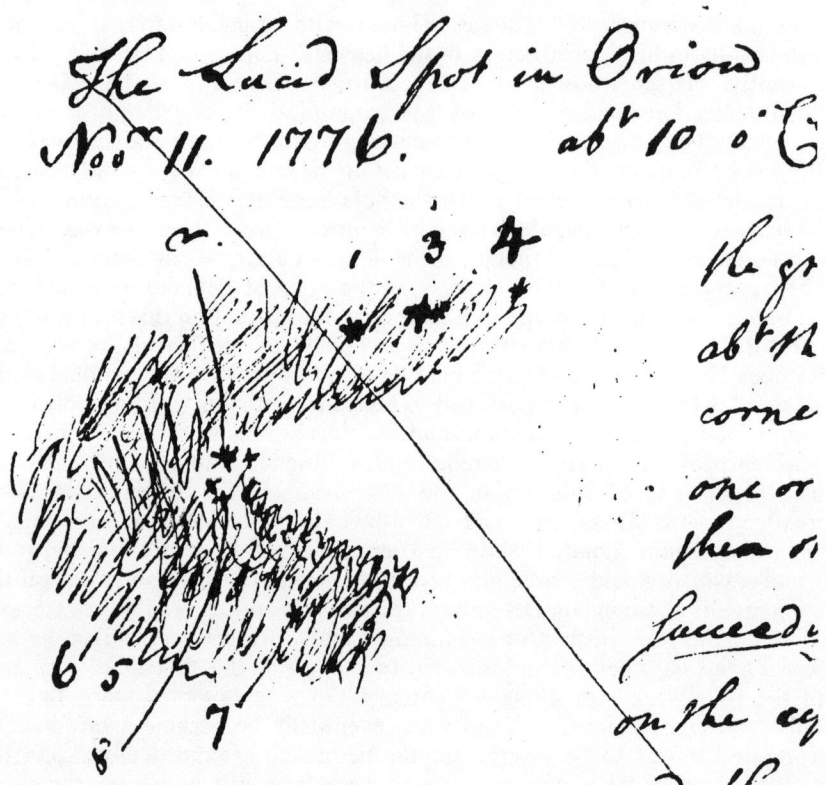

FIG. 2. Herschel's sketch of the Orion Nebula, M42, made on 11 November 1776 (W.2/1.1, f. 37).

investigation of their nature, is dominated by two questions of observation: Are nebulae 'resolved' into stars by large telescopes?, and, Do nebulae change? This is because nebulae resolved into stars are (if the stars are not illusory) star systems; on the other hand, nebulae that change perceptibly in a few short decades must be small and so *cannot* be of the size of star systems. It would be no less than five years before Herschel looked at any nebula other than that of Orion, but during those years he did examine the Orion Nebula seven more times[16] and took two more sketches of it (Figures 2 and 3). And on 15 December 1788, although he found the handful of isolated stars in the nebula positioned as before, he remarked: "But there is a visible alteration in the figure of the lucid part."[17]

Between the autumn of 1779 and the arrival of the Messier catalogue from Watson in December 1781, Herschel enlarged his list of observed nebulae from one to four.[18] In addition to examining the Orion Nebula on some seven occasions, he looked at three more of Halley's list of six nebulae: the globular cluster M13 (observed once, as a "nebula without stars in it"[19]), the galactic cluster M11 (observed four times, typically described as "an amazing multitude of small stars"[20]), and the Andromeda Nebula M31 (observed twice: "Has no

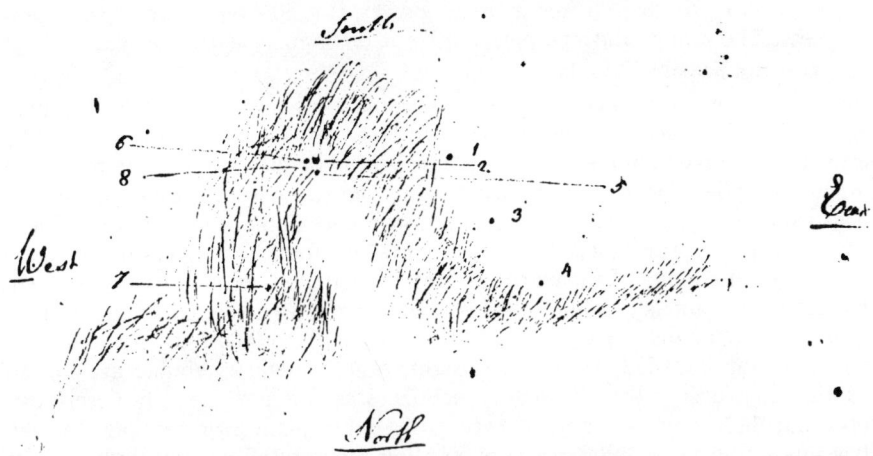

FIG. 3. Herschel's sketch of the Orion Nebula, M42, made on 25 January 1778 (W.2/1.1, f. 47).

star in it"; "no star visible"[21]). But his interest had been more thoroughly aroused than this brief list suggests, for in his *Journal* for 6 August 1780 he had listed objects "To be observed", notable among them "All the Nebula's, their stars counted, and the form delineated".[22] Nevertheless, and notwithstanding the wealth of new nebulae announced in Messier's list and the opportunity they presented to Herschel's 20ft reflector with its 12inch mirrors, Herschel seems not to have looked at a single nebula after July 1781 until his arrival at his new home near Windsor Castle on 2 August 1782.

Now that he was a professional astronomer, Herschel began at last to take a more serious interest in nebulae, though still in intermittent fashion. On 5 August he came across a "nebulous star or telescopic comet" (which proved to be M5),[23] and this no doubt prompted him to set Caroline to sweep for comets with a small refractor in imitation of Messier. On 29 August Herschel observed two more Messier nebulae: M52 ("a group of small stars"), and the Ring Nebula in Lyra, M57 ("Extremely curious not quite round. The greatest light is at the outside and in the middle it seems to be dark. I suspect the left side at least to consist of very small stars.").[24] On 30 August Herschel showed M31 and M11 to the King,[25] proof that he was by now conscious of nebulae as interesting objects for his large telescopes.

Early in September he examined M15 ("no star visible"[26]) and M31 ("no star"[27]), but failed to find M26; and on 7 September he made his first major discovery among the nebulae:

> A curious Nebula. or what else to call it I do not know. it is of a shape somewhat oval, nearly circular, and with this power [460] appears to be about 10 or 15" diameter. It is of the same shape with 278 [magnification] but much less in appearance. with 932 it is of still the same shape but much larger. So that its appearance seems to follow the law of magnifying, from whence it is clear that it is of some real magnitude in the heavens and not a glare of light. The brightness in all the powers does not differ so much as if it were of a planetary nature, but seems to be of the starry

kind, tho' no star is visible with any power. It is all over of the same brightness. The compound eye piece will not distinguish it from a fixt star, at least not sensibly....[28]

This momentous discovery of what has become known as the "Saturn Nebula" was to have far-reaching consequences for Herschel. It was the first of what he termed 'planetary nebulae', objects whose status in the cosmos was to puzzle him for decades. He was to return to this planetary nebula time and again, measuring its position with extreme care in a check for proper motion,[29] and it became a favourite object for him to show visiting astronomers in the years ahead—proof enough of the worrying ambiguity he sensed in its nature. Indeed, the status of planetary nebulae in his theorizing shifted several times; and we shall not understand his cosmological papers of the 1780s until we close our minds to our knowledge, from hindsight, that planetary nebulae are indeed nebulae, and accept that 'planetary nebula' was for him a purely descriptive term and these objects were not necessarily either planets or nebulae. Indeed, they might not be assimilated to any existing category of celestial body.

Still Herschel hesitated to apply himself seriously to the study of nebulae, though he had further confirmation of changes in the Orion Nebula when on 31 January 1783 he noted that "the nebulous part is quite different from what it was last year. The 9th star very strong, the nebula about it and the 8th being much dispersed".[30] But meanwhile Caroline in sweeping for comets with her primitive refractor was scoring notable successes among the nebulae. On 30 September 1782 she came across M27, which her brother had not yet seen,[31] and on 26 February 1783 she discovered two previously-unknown nebulae.[32] She found two more on 4 March, and a cluster on 8 March, while on 7 April she came across M56.[33]

No doubt impressed, Herschel took time off from his self-imposed task of examining every star listed by Flamsteed and on 4 March "began to sweep the heaven for nebulas and clusters of stars. I chose for this purpose the 3½ft achromatic with a single eye-lens. The field I suppose is near 50'... and there is an admirable quantity of good light in the center".[34] These sweeps were soon broken off; but in May there are nights when he observes two or three Messier nebulae, and suddenly, on the three nights of 30 and 31 July and 2 August 1783, a year and a half after receiving the 1780 Messier catalogue, we find him observing a total of no fewer than 23 Messier objects.[35]

Herschel was now in some confusion as to the nature of the nebulae. On the one hand, the only nebula he had repeatedly examined over the years, the Orion Nebula, had (he believed) altered shape, and it was continuing to do so: on 20 September he noted it was "changed"[36] and on 28 September it was "surprizingly changed".[37] It therefore could not possibly be a vast star system, and Herschel had recognised this, remarking of M82 on 6 August: "... whether the light which is also extended consists of stars *or nebulosity such as in Orion's Sword handle* I can not resolve" (italics supplied).[38]

On the other hand, he several times speaks of nebulae and clusters as though he sees them as members of a single continuous series. For example:

> [M9, 3 May 1783] I see several stars in it & make no doubt a higher power & more light would resolve it all into stars. This seems to be a good nebula

for the purpose of establishing the connection between Nebulas in general & stars in clusters.[39]

[M17, 31 July 1783] A very singular Nebula; it seems to be the link to join the Nebula in Orion to others for this is not without a possibility of being stars.[40]

[M30, 21 August 1783] Plainly resolved into very small stars. It is a difficult step i.e. if we divide the transition from the Pleiades down to the Nebula in Orion into six steps this is perhaps the 4th towards the real nebulas.[41]

On 7 August 1783 he asked himself whether nebulous appearances might not result simply from light sources (stars) insufficiently powerful to make a clear impression on the eye:

Relating to the appearance of Nebulae without stars. 1 tried some excessively small stars near ν Aquilae; when ν was perfectly distinct & round the *extremely* small stars were dusky or not perfectly defined; (these stars I have often found in my 20ft reflector appear defined & without that duskyness; which difference I ascribe to want of light in one case and a sufficiency in the other to make an impression of a round point.) The *excessively* small stars were still less defined; and as there are in this neighbourhood stars of all sizes, I saw some so very small that they gave the idea of a small dusky spot, approaching to a nebulous appearance. By very long attention I perceived small dusky nebulous spots which without such attention I perceived might be in view without the least suspicion.[42]

Fortunately Herschel was just completing his fine new 20ft reflector, with a mirror of 18inches diameter and a reliable and stable mounting—the instrument with which he and his son John were to do their finest work. Instead of rapidly sweeping for nebulae with low-powered telescopes in a manner more suited to the swift detection of newly-arriving comets, as he had briefly done before, Herschel decided to devote this uniquely-powerful reflector (and much of the next two decades of his life) to methodical and very detailed sweeps of the sky, section by section, in the hunt for nebulae. On 23 October 1783 the new telescope was used for the first time; on the 28th, Herschel made a trial sweep; and on the 29th, the numbered sweeps began.

In addition to yielding new nebulae in vast numbers, Herschel's sweeps were to make him very experienced in the distribution of nebulae across the sky, and this had important implications for his theories of the nature of nebulae. For example, on 30 December 1783 he commented:

It appeared to me remarkable that in and about the place where the many Nebulas began there was an uncommon scarcity of stars so that many fields were totally without a single star. If these Nebulae should be clusters of stars it should seem as if they were collected together from the neighbouring spaces; however such a surmise as this will want a number of facts to give it any sort of consequence: it is a mere *appercu* [sic] which do as often miss as they hit.[43]

The Milky Way he envisaged as a (compound) layer or stratum of stars of which the Sun is one, and he soon began to explain the far-from-uniform distribution of the nebulae across the sky by arguing that they likewise are gathered into strata, presumably by the power of gravity or similar forces. These insights he embodied in the epoch-making (if ill-organised) paper from which we have already quoted, "Account of some observations tending to investigate the construction of the heavens", which was read to the Royal Society on 17 June 1784.[44] Its most striking feature is the explanation of how star counts might be utilized to establish the distance to the borders of the Milky Way in a given direction; but Herschel also outlines his current programme of sweeps for nebulae, makes the misleading claim concerning resolution of Messier nebulae that we discussed earlier, and lists some of the wonders of the nebulae. These include nebulae

> of the cometic shape, with a seeming nucleus in the center; or like cloudy stars, surrounded with a nebulous atmosphere; a different sort again contain a nebulosity of the milky kind, like that wonderful, inexplicable phaenomenon about θ Orionis; while others shine with a fainter, mottled kind of light, which denotes their being resolvable into stars.[45]

In spite of this sentence, it has been usual to read this paper as already embodying Herschel's later belief that nebulae are merely star clusters disguised by distance, a misunderstanding fostered by his claim to have resolved most of the Messier nebulae he had examined. In fact Herschel is here hinting at how he distinguishes true nebulae from star clusters disguised by distance: true nebulae have a milky appearance, whereas distant star clusters betray their nature by the appearance of mottled nebulosity which he usually termed 'resolvable'.

On 22 June 1784, just five days after this paper was read to the Royal Society, Herschel came across M17, the so-called "Omega" Nebula. It provided the missing link between the milky nebulosity of the supposedly true nebula and the resolvable nebulosity of the distant star cluster; and the presence of *both* together in the extensive Omega Nebula suggested that there was perhaps no difference *in kind* between the two nebulosities, but rather that star clusters would appear as mottled, resolvable nebulosity at large distances, and as milky nebulosity at greater distances still.

> A wonderful Nebula. Very much extended, with a hook on the preceding side; the nebulosity of the milky kind; several stars visible in it, but they seem to have no connection with the Nebula which is probably far more distant. I saw it only thro' short intervals of flying clouds and hazyness; but the extent of the light including the hook is about 10 minutes. I suspect besides that on the following side it goes on much farther and diffuses itself towards the north & south. It is not of equal brightness throughout, and has one or more places, where the milky nebulosity seems to degenerate into the resolvable kind. Such a one is that just following the hook towards the north. Should this be confirmed on a very fine night, it would bring on the step between these two nebulosities which is at present wanting, and would lead us to surmize that this nebula is a stupendous Stratum of immensely distant fixed stars some of whose branches are near enough to us to be visible as resolvable nebulosity, while the rest runs on to so great a distance as only to appear under the milky form.[46]

Confirmation came on 19 July, when Herschel examined the "Dumbell Nebula", M27. This

I suppose to be a double stratum of stars of a very great extent. The ends next to us are not only resolvable nebulosity but I really do see very many of the stars mixt with the resolvable nebulosity. Farther on the nebulosity is but rarely resolvable & ends at last in milky whitishness of the same appearance as that in Orion. The Idea I form of the shape of the strata is

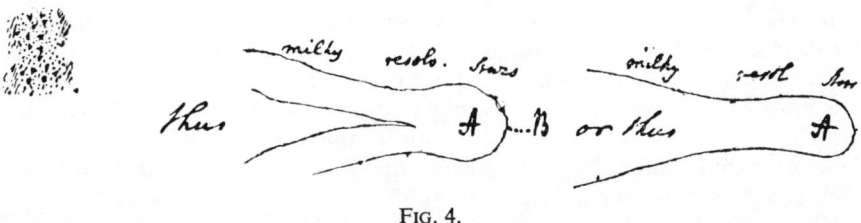

FIG. 4.

[see Figure 4]. These two being laid on each other, A on A and viewed from B, so as to have the small round end A foremost may produce the appearance of this curious nebula.[47]

Herschel accordingly concluded that all nebulae (which term did *not* necessarily include planetary nebulae) were star clusters more-or-less successfully disguised by distance. And since clusters encouraged him to think of a clustering power (such as gravity) at work over long periods of time,[48] he was naturally led in his second (1785) paper on the construction of the heavens,[49] to propose a cosmogony in which the universe began with stars scattered throughout infinite space, and these under the action of gravity are collecting into strata such as those that form the Milky Way, which strata are in time fragmenting into smaller and more compressed clusters. Resolutely closing his mind to the changes he had so frequently observed in the Orion Nebula (and which he would cite in evidence in later years[50]), Herschel declares it to be a galaxy comparable to our own.

Nevertheless, Herschel concludes what would otherwise be a starkly crisp cosmogony by listing planetary nebulae, "that from their singular appearance leave me almost in doubt where to class them".

> The planetary appearance of the first two is so remarkable, that we can hardly suppose them to be nebulae; their light is so uniform, as well as vivid, the diameters so small and well defined, as to make it almost improbable they should belong to that species of bodies. On the other hand, the effect of different powers seems to be much against their light's being of a planetary nature, since it preserves its brightness nearly in the same manner as the stars do in similar trials. If we would suppose them to be single stars with large diameters we shall find it difficult to account for their not being brighter. . . . We might suspect them to be comets about their aphelion, if the brightness as well as magnitude of the diameters did not oppose this idea; so that after all, we can hardly find any hypothesis so probable as that of their being Nebulae [and therefore consisting of stars]; but then they must consist of stars that are compressed and accumulated in the highest degree. If it were not perhaps too hazardous to pursue a

former surmise of a renewal in what I figuratively called the Laboratories of the universe, the stars forming these extraordinary nebulae, by some decay or waste of nature, being no longer fit for their former purposes, and having their projectile forces, if any such they had, retarded in each others' atmosphere, may rush at last together, and either in succession, or by one general tremendous shock, unite into a new body.[51]

This, he suggests, might account for Tycho's nova of 1572.

Herschel's third (1789) paper on the construction of the heavens (in the form of Introductory Remarks to his second catalogue of a thousand new nebulae and clusters of stars) consists of reflections on the cosmogony already presented, with hints of how repulsive forces might combine with gravity to bring each star cluster eventually into globular form, the cluster finally becoming a planetary nebula which "may be looked upon as very aged, and drawing on towards a period of change, or dissolution".[52]

Although in these 1785 and 1789 papers Herschel had suggested that planetary nebulae may be star clusters in the last stage of their evolution, the planetaries continued to worry him and to fascinate a succession of astronomers visiting his observatory.[53] Indeed, some weeks after the publication of the 1785 paper he wrote to Lalande describing planetaries as "des corps celèstes dont nous n'avons pas encore d'idée bien claire & qui sont peut-être d'un genre tout à fait different de ce que nous connoissons dans les cieux".[54] In other words, the cosmogony of his 1785 and 1789 papers did not have the simplicity we often attribute to it, for the planetaries fitted uneasily into the cosmogony, in their rôle as clusters on the verge of dissolution. It may therefore have been with some relief that in November 1790 he came across a planetary nebula with a visible central star (the two together classified by him as a 'nebulous star'). It was, he recorded, "a most singular phaenomenon! A star of about the 8th magnitude, with a faint luminous atmosphere"[55] and, as is well known, this convinced him once more of the existence of true nebulosity: the star, he thought, was condensing out of the nebulosity under gravity. One can almost sense the relief as he incorporates 'nebulous stars', 'planetary nebulae with centres', and 'planetary nebulae' as nebulous, *pre*-stellar stages in an extended and revised cosmogony, rather than as impending final catastrophes—relief, too, as he can accept the Orion Nebula once more as a true nebula, close at hand, small, and liable to rapid alteration.[56]

The period during which Herschel equated all nebulae with star clusters, then, was brief: from the summer of 1784 (and *after* the presentation of his 1784 "contruction of the heavens" paper) until he saw the "most singular phenomenon" in November 1790. And even during this period, planetary nebulae offered worrying resistance to the straightjacket of his purely stellar cosmogony.

However, it is for the decade 1774-84 that the main revisions to the accepted story must be made. Herschel understood from the outset the great significance that would attach to observed changes in a nebula such as that of Orion; this one nebula he did observe repeatedly, and he satisfied himself that it had indeed changed. Otherwise his investigations of nebulae did not gather momentum until the summer of 1783—long after he had received the 1780 Messier catalogue, and only a few weeks prior to the commissioning of his fine "large"

20ft reflector. In the confusing situation in which he soon found himself, Herschel believed he could distinguish visually between the 'mottled' or 'resolvable' nebulosity of star systems and the 'milky' nebulosity of true nebulae. But in June and July 1784 he came across two nebulae in which the two nebulosities were present together, suggesting there was no real distinction between them; and it was then that he adopted the cosmogony of his 1785 and 1789 "construction" papers in which nebulae are equated with star clusters.

Acknowledgements

I am grateful to the Royal Astronomical Society for permission to quote from manuscripts in their possession and to reproduce four drawings by Herschel.

REFERENCES

1. W. Herschel, "Account of some observations tending to investigate the construction of the heavens", *Philosophical transactions*, lxxiv (1784), 437–51, pp. 439–40. Herschel refers to Messier's second catalogue, published in 1780 in the *Connoissance des temps* for 1783 and containing M1 to M68. Messier's final catalogue appeared in the same periodical the following year, and Dr D. W. Dewhirst points out that imperfections in the type show that standing type from the second catalogue was used in the third after correction.
2. Letter of William Watson Jr to Herschel, 7 December 1781 (Royal Astronomical Society Herschel MSS, 13.W.11).
3. Herschel MSS, W.4/23.
4. On Herschel's telescopes, see J. A. Bennett, " 'On the power of penetrating into space': The telescopes of William Herschel", *Journal for the history of astronomy*, vii (1976), 75–108.
5. W. Herschel, "On the construction of the heavens", *Philosophical transactions*, lxxv (1785), 213–66; "Catalogue of a second thousand of new nebulae and clusters of stars; with a few introductory remarks on the construction of the heavens", *ibid.*, lxxix (1789), 212–55.
6. J. A. Bennett, "Catalogue of the archives and manuscripts of the Royal Astronomical Society", *Memoirs of the Royal Astronomical Society*, lxxxv (1978).
7. See for example: Mrs John Herschel, *Memoir and correspondence of Caroline Herschel* (2nd edn, London, 1879), chap. 2.
8. W. Herschel, "On the parallax of the fixed stars", *Philosophical transactions*, lxxii (1782), 82–111.
9. W. H. Steavenson, "Some eye-pieces made by Sir William Herschel", *Monthly notices of the Royal Astronomical Society*, lxxxiv (1923–24), 607–10.
10. Herschel was deemed "fit for Bedlam" (a reference to the famous lunatic asylum), as reported by William Watson in his letter of 18 December 1781 (Herschel MSS, 13.W.12). Alexander Aubert offered to accompany him there (letter of 22 January 1782; 13.A.8).
11. Herschel MSS, W.2/1.4.
12. Herschel MSS, W.2/1.1.
13. Robert Smith, *A compleat system of opticks* (Cambridge, 1738), ii, 447-8.
14. Herschel MSS, W.2/1.1, f. 1.
15. *Ibid.*
16. On 9 April 1774, 11 November 1776, 25 and 26 January 1778, 7 and 25 February 1778, and 15 December 1778.
17. Herschel MSS, W.4/1.1, f. 7.
18. In addition to a Milky Way cluster observed on 24 August 1780.
19. On 22 August 1779, W.4/1.1, f. 8; this is possibly Herschel's first use of the term 'nebula'.
20. On 25 June 1781, W.4/1.2, f. 106.
21. On 6 August 1780, W.4/1.1, f. 63, and 21 July 1781, W.4/1.2, f. 112.

22. W.2/1.2, f. 4r.
23. W.4/1.3, f. 217.
24. Ibid., f. 223.
25. Ibid., f. 224.
26. On 3 September 1782; ibid., f. 226.
27. On 4 September 1782; ibid., f. 226.
28. Ibid., f. 231.
29. For example, in 1783 on 30 July, 25 August, 20 September, 17 and 23 October and 14 November, and in 1784 on 28 June and 15 July.
30. W.4/1.4, f. 319.
31. W.4/1.3, f. 240.
32. W.4/1.4, f. 333.
33. Ibid., ff. 338, 344, 370.
34. Ibid., f. 338.
35. W.4/1.5, ff. 398–404.
36. Ibid., f. 432.
37. Ibid., f. 441.
38. Ibid., f. 405.
39. W.4/1.4, f. 379.
40. W.4/1.5, f. 400.
41. Ibid., f. 414.
42. Ibid., f. 406.
43. Ibid., f. 484–5.
44. See ref. 1.
45. Ibid., 443.
46. W.4/1.7, ff. 642–3.
47. W.2/1.9, f. 29r.
48. For example, on 17 July 1784 he commented of a "cluster of not very compressed stars" that it "may be compared to a cluster which is forming or gathering and not yet arrived to the state of those that are more advanced, or contain more stars" (W.2/1.9, f. 27v).
49. See ref. 5.
50. "The changes I have observed in the great milky nebulosity of Orion, 23 years ago, and which have also been noticed by other astronomers, cannot permit us to look upon this phenomenon as arising from immensely distant regions of fixed stars" ("Catalogue of 500 new nebulae, nebulous stars, planetary nebulae, and clusters of stars; with remarks on the construction of the heavens", *Philosophical transactions*, xcii (1802), 447–528, p. 499).
51. Pp. 265–6.
52. *Op. cit.* (ref. 5, 1789), 225.
53. For example, on 27 November 1787, Herschel showed his visitors the planetary nebula H.IV.18: "Messrs Cassini, Mechain Le Seure & Carochet saw this nebula, and the moon being absent, it appeared in its usual planetary view; these Gentlemen saw it very well and admired it as a great curiosity. Mr Cassini observed that a very small fixt star nf the nebula appeared not unlike a sattelite to it" (W.2/3.7). Planetary nebulae were also shown to Dr Watson and a Mr Marsden on 11 March 1788, to Lord Palmerston on 3 August 1788, to Lalande on 5 August 1788, he "having never before seen a planetary one", to Edward Pigott on 6 August 1788, and to the Abbé Ximenes on 13 April 1789.
54. W.1/1, f. 131.
55. W. Herschel, "On nebulous stars, properly so called", *Philosophical transactions*, lxxxi (1791), 71–88, p. 82.
56. In his 1802 paper (ref. 50).

2. *The Nebulae from Herschel to Huggins*

We have seen how Herschel's early years as an astronomer saw shifts in his theories of the nature of the nebulae. When he arrived near Windsor Castle in 1782, he was equipped with a 20ft reflector of substantial mirror size (12 inches) but with rickety mounting (Figure 1). He then had little familiarity with nebulae, but he had examined the Orion Nebula sufficiently to be convinced that it had altered shape, and therefore must be a true nebula (whatever that might mean) rather than a star cluster disguised by distance. A year later, with the completion of his 'large' 20ft with its 18-inch mirrors and excellent mounting (Figure 2), he was able to begin a systematic search for specimens of nebulae, of which he eventually accumulated two and a half thousand. His work was, as we shall see, revised and extended to the southern skies by his son John, so that in 1847 F. G. W. Struve could write: "L'étude de ciel nébuleux paraît être le domaine presque exclusif des Herschel."[1]

In that sentence we have a sign of the isolation of Herschel from the astronomical community that makes of him something of the tragic hero. He had shown his lack of training and expertise, for example, by the inept way in which he defined the position of his 'comet' (Uranus) and by his blunders over its parallax.[2] He had mixed speculations about lunar inhabitants with calculations of the height of lunar mountains.[3] He had casually mentioned eyepieces with magnifications of thousands as though this was commonplace. Small wonder that some at the Royal Society thought him "fit for Bedlam".[4]

Yet it could not be denied that in the course of searching the sky for double stars this organist had somehow discovered a major new planet, recognised by him at a glance as no ordinary star. His subsequent journey to Greenwich where he set up one of his telescopes alongside the Observatory instruments convinced the Astronomer Royal of his quality as a telescope maker; and his ruthless devotion to observing in all weathers, and his experience in coaxing the best results from temperamental instruments, served to confirm his isolation from other astronomers who were without access to huge, cosmological telescopes. While they were dissecting problems of the solar system with a scalpel, Herschel was laying about him with a stellar broadsword. "Comme personne n'avoit, ou ne croyoit avoir d'instrument assez fort pour distinguer ces merveilles", wrote J.-D. Cassini in 1784, "il fallut y croire sur sa parole".[5] The completion in 1789 of his monster 40ft reflector (Figure 3) with mirrors of 4ft diameter weighing about a ton confirmed his ascendancy. It became one of the wonders of the age, visited by the King and the Archbishop of Canterbury (although in fact the mounting was to prove too cumbersome and the mirrors tarnished quickly).[6] "M. Herschel ... nous assure qu'à l'aide des grossissemens étonnans que ses téléscopes supportent, il est parvenu à voir...", wrote F. T. Schubert,[7] thereby expressing the impossibility of other astronomers coming to a considered judgement of his work: they could not judge him as peers because they could not see what he saw.

Nor did they share his interests or understand his methods. He was collecting

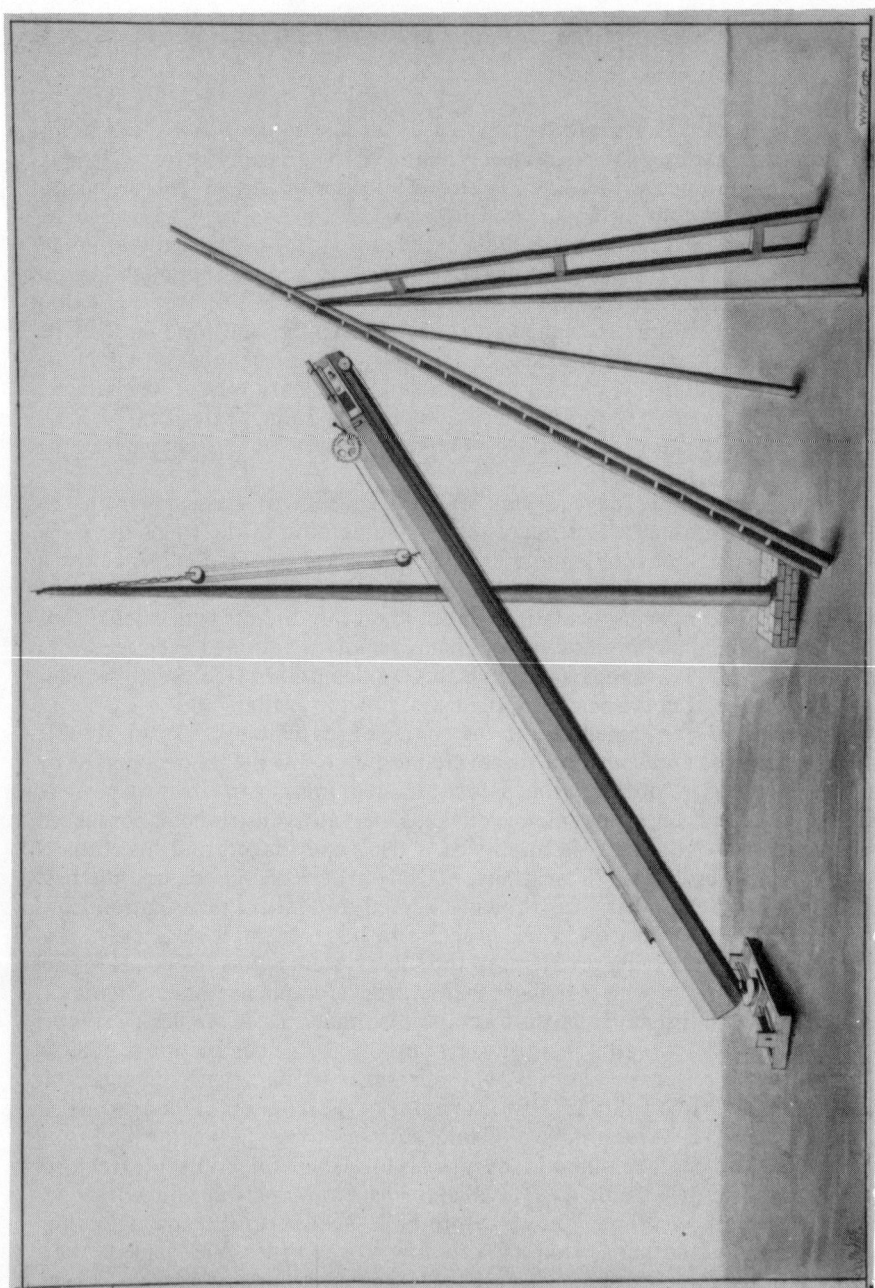

FIG. 1. William Herschel's 'small' 20ft reflector. From a drawing by William Watson, RAS Herschel MSS W.5/5, no. 4.

double stars by the hundred and nebulae by the thousand and arranging them in species, artificial or natural, as though astronomy were somehow a branch of natural history;[8] the astronomy they knew was above all the study of the planets, satellites and comets of the solar system, each of which was an individual with known and often unique characteristics and subject to mathematical laws. Herschel's theories of the life-cycle of nebulae could not be judged by accepted norms, since no astronomer had ever attempted this kind of astronomy. Throughout his career Herschel was to be the subject of envious attacks – "Those who envy you for such there are", wrote his ally William Watson on the completion of the 40ft, "will sicken at the news, & will redouble their efforts in depreciating your merits"[9] – but even Watson was often left baffled and anxious. There was no way in which the astronomical community could repeat his observations, or measure his natural history against accepted professional models. As a result, Herschel was in large measure writing for future generations, though the readiness with which the pages of *Philosophical transactions* were made available to him ensured that the concepts and methods he created were indeed accessible to his successors.

We have seen in chap. 1 that until 1784 Herschel distinguished between true nebulae with their milky appearance, and distant star clusters which appeared mottled; but that in the summer of that year he encountered nebulae in which both milky and mottled nebulosity were present, and he now interpreted the different appearances in nebulosity as the result of the different *distances* at which the stars forming the 'nebulosity' lay from the observer. If he did not go so far as to declare categorically that all nebulae were star clusters, it was because a natural historian could not exclude the future discovery of new species, and in any case his planetary nebulae were perplexing. But central to his 1785 and 1789 papers is the vision of scattered stars gradually coming together under an attractive force such as gravity, and forming increasingly condensed star clusters many of which appear to us as nebulae. This being so, an extensive nebula like the Orion Nebula or the Andromeda Nebula, which was nevertheless so distant as to appear milky, must contain huge numbers of stars and may well "outvie our milky-way in grandeur"[10] – in other words, be a true galaxy.

He was however unhappy about the catastrophes that seemed the inevitable end result of this development in time, though he thought that the counter-attraction of innumerable stars, the existence of orbital motions, and perhaps even repulsive forces, would all contribute to postpone gravitational collapse.[11] These problems were dramatically reduced by the discovery of the 'nebulous star' in November 1790, consisting of a central star with an extensive surrounding shell of nebulosity out of which it was apparently condensing.[12] Recognising that 'true' nebulosity existed after all, Herschel envisaged nebulous stars, planetary nebulae and the like as intermediate stages between the earlier and purely nebulous period and the later and purely stellar period in the life-history of a nebula/cluster. In his old age, he drew on the long catalogues of nebulae that he had compiled, and paraded for the reader examples of very diffuse nebulosity, of more condensed nebulosity, of nebulosity out of which stars are beginning to form, and so on, ending with highly condensed star clusters (see Figure 4). These, he claimed, were specimens arranged by age just as

Fig. 2. William Herschel's 'large' 20ft reflector, from an engraving published by him in February 1794.

a naturalist might teach the life-history of a plant by pointing out specimens at successive stages of development.[13]

It was a methodological step so bold that even his own son John would not be able to follow him. How, John was to ask in 1826, after his father's death, is one to make the step from "observed graduation" – which William had indeed shown – to "concluding them [the nebulae and star clusters] to be in a course of progress from one state in the series to another"? "So wide is the field of conjecture, and so uncertain the analogies we have to guide us, that we shall do well for the present to dismiss hypothesis, and have recourse (perhaps for centuries to come) to observation."[14]

But William Herschel's speculations did not end here. Where did the initial nebulosity come from? Perhaps from the light given off by stars and nebulae, that must be present throughout space and, here and there, be dense enough to exist as nebulosity.[15] Furthermore, small clouds of nebulosity will sometimes be attracted towards a star in the form of a comet, some of whose material will fall

FIG. 3. William Herschel's 40ft reflector, from an engraving published by him in February 1794.

into the star and replenish it; at the same time the force of the star's heat and light will consolidate some of the material to give the comet a solid nucleus, and this, after numerous such passages past stars, may become large enough to be a planet. Indeed a star is itself nothing more than a large (and inhabited) planet with an outer atmosphere of luminous clouds.[16]

And so each region of Herschel's universe is the scene of endless cycles in which light has a fundamental role and gravity is the great agent of change in time. He is the unique example of a fine telescope builder and a great observer, who nevertheless transcends the observations he has made with his own telescopes and who declares publicly his intention to speculate too much rather than too little. Small wonder that in 1820, when Herschel was in his eighties, the future Royal Astronomical Society (of which he would be first President) should write sceptically of his theories in the Address "explanatory of their views and objects":

> Beyond the limits however of our own system, all at present is obscurity. Some vast and general views of the construction of the heavens, and the laws which may regulate the formation and motions of sidereal systems, have, it is true, been struck out; but, like the theories of geologists, they remain to be supported or refuted by the slow accumulation of a mass of facts....[17]

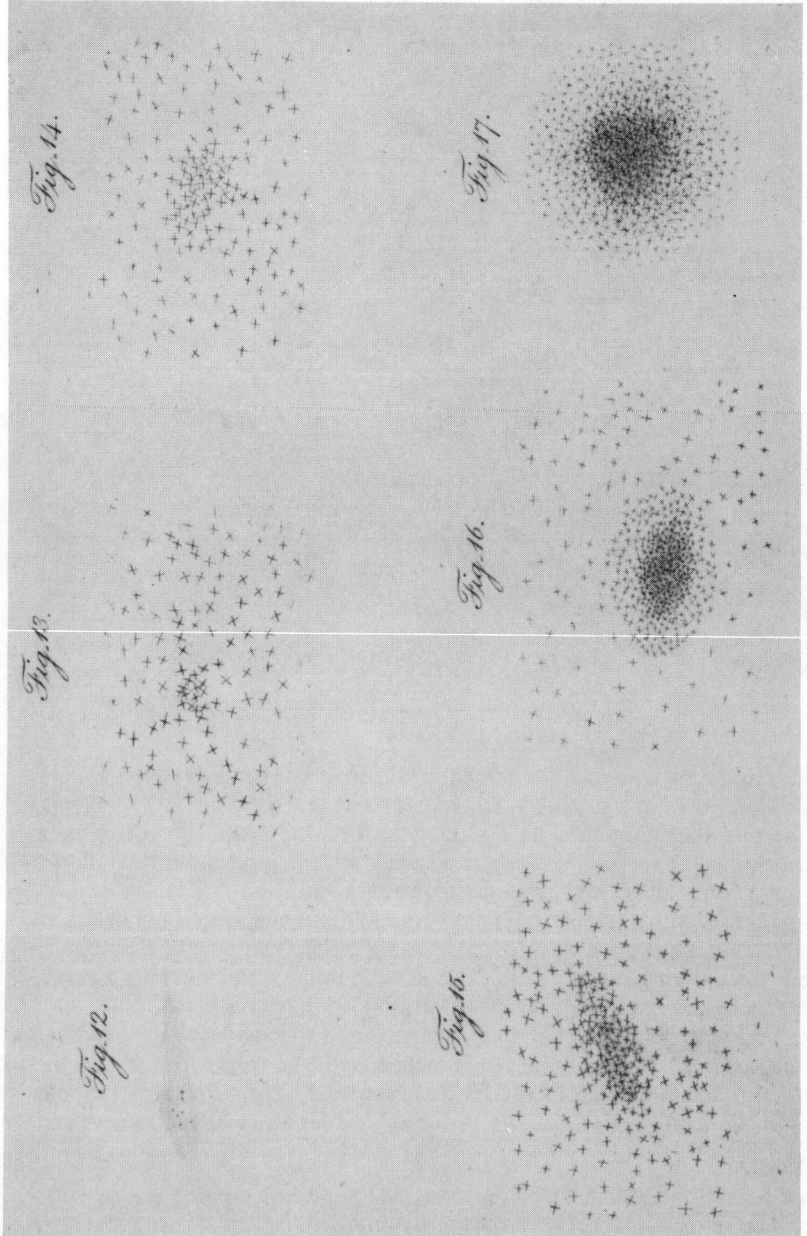

FIG. 4. The later stages of the life-cycle of a star cluster as it condenses under the action of gravity, from Herschel's 1814 paper: "...there is perhaps not so much difference between them ... as there would be in the annual description of the human figure...."[13]

John Herschel offers an interesting contrast to his father.[18] Born in 1792, when his father was aged 53, his early manhood was the time of his father's old age. Whereas William had been brought up in poverty and obscurity, John's mother had been a rich widow before she married William, and his father was world-famous; whereas William was self-taught and had both the limitations and the sturdy independence of the autodidact, John had a conventional education and found himself one of a talented generation in Cambridge; whereas William had come eagerly to astronomy in middle life and "worked as though he had few tomorrows", John in his early twenties had to be coerced away from the university career on which he had set his heart so that William's wealth of experience could be handed on before it was too late. John's gifts were equal to those of his father, and he was to become a central figure of Victorian scientific society (and Master of the Mint). He was an apt apprentice as a telescope maker, and a fine observer. But his judgements were more considered, he could not close his eyes to contrary evidence, and his theories were advanced tentatively and always subject to correction by new observations. With the refurbished 20ft, John resurveyed his father's nebulae, and then, in 1834, took the telescope to the Cape of Good Hope for four years to sweep the southern skies. His "Catalogue of nebulae and clusters of stars" (1864) contains over five thousand objects, and was to be the basis of the New General Catalogue used today.

It was perhaps necessary that William's hectic theorising should be followed by a period of consolidation, and for two decades after William's death in 1822, his belief in 'true nebulosity' was unchallenged. George Biddel Airy spoke for all when he described the existence in the sky of "nebulous matter in the wildest confusion", and remarked of the Orion and Andromeda Nebulae: "No one ... who has seen these in a telescope of great light ... can persuade himself that these can be anything but masses of nebulous matter."[19] But in 1839, at Birr Castle in central Ireland, the future third Earl of Rosse (then Lord Oxmantown) completed a reflector with 3ft mirrors and mounted in the manner of Herschel's 20ft (see Figure 5).[20] Although nominally of less power than the 40ft with its 4ft mirrors, it was much more effective in use, and considerably larger than the 20ft. Indeed, its only true rival had been the fine 25ft reflector with 3ft mirrors that William Herschel had made for the King of Spain (see Figure 6), but this had been destroyed by Napoleonic troops.[21]

The purpose behind the Rosse telescope was the answering of the age-old question of whether 'true nebulosity' exists, a question all the more important in the light of Laplace's nebular theory of the origin of the solar system. Rosse himself showed exemplary caution in drawing conclusions from his observations. Although he was able, as he thought, to announce the resolution into stars of several of the most notable nebulae, he warned his readers in 1844 that "still it would be very unsafe to conclude that such will always be the case, and thence to draw the obvious inference that all nebulosity is but the glare of stars too remote to be separated by the utmost power of our instruments".[22]

But no such circumspection marked the pronouncements of his redoubtable adviser, Dr T. R. Robinson, Director of the Armagh Observatory for nearly sixty years. As early as 7 November 1840, Robinson recorded:

I hope that Lord O may succeed in passing the magnitude of this telescope;

but even should he fail in this, I am certain that one [whose mirror is constructed] in pieces, but with fewer divisions than N° 1 would be inestimable for Nebulae both in respect of adding to our knowledge, and of rectifying received opinions, for I confess that the apparent resolvability [into stars] of the two [nebulae] I examined has made me doubtful of the existence of that assumed nebulous matter which is the basis of much speculation respecting the formation of stars and planets.[23]

1845 saw the completion at Birr Castle of the "Leviathan of Parsonstown", a

FIG. 5. A contemporary photograph of the Birr reflector with 3ft mirror; the skeletal tube replaced the original conventional tube. Note the obvious resemblance of the mounting to that of Herschel's 20ft. (Royal Astronomical Society archives.)

monster reflector with mirrors of 6ft diameter, mounted in a tube facing south and supported between two huge walls (Figure 7). How William Herschel would have approved of this crude but effective escalation of the scale of cosmological telescopes, by another self-taught amateur! Robinson's private account of the first observations concludes in characteristic style: "We have however worked to some purpose, for of the 43 nebulae which we have examined *All have been Resolved*.... In fine, this instrument bids fair to throw a light hitherto not merely unattainable, but unhoped, on the constitution of the sidereal universe."[24] In public he did not hesitate to draw conclusions:

> [Robinson] could not leave this part of his subject without calling attention to the fact that no real nebula seemed to exist among so many of these objects chosen without any bias: all appeared to be clusters of stars.... If it prove to be the case that *all* the brighter nebulae yield to this telescope, it appears unphilosophical not to make universal Sir J. Herschel's proposition, that "a nebula, at least in the generality of cases, is nothing more than a cluster of discrete stars".[25]

And in 1848:

> Above fifty nebulae, selected from Sir John Herschel's catalogue, without any limitation of choice but their brightness, were *all resolved without exception*. From this he conceives himself authorized to ask, is there any evidence that nebulous matter has real existence?[26]

Just as William Herschel had been deceived in many of the nebulae he once thought he had resolved, so too were Lord Rosse and his colleagues. But the great nebula in Orion still remained as the outstanding challenge: John Herschel had reported that its appearance was "very different from what might be supposed to arise from the congregation of an immense collection of small stars".[27] J. P. Nichol, professor of astronomy in the University of Glasgow and until then a noted exponent of a nebular theory, recounts in the later editions of his *Architecture of the heavens* the excitement with which he and Lord Rosse at Christmas 1845 turned the great telescope to Orion for the first time. "With an anxiety natural and profound", he writes, "the scientific world watched the examination of Orion by the six-feet mirror: for the result had either to confirm Herschel's hypothesis, in so far as human insight ever could confirm it – or unfold, among the stellar groups, a variety of constitution not indicated by those in the neighbourhood of our galaxy". Alas, the Irish climate was unfavourable: "Not yet the veriest trace of a star."

But the disappointment was only temporary. On 19 March 1846, Rosse wrote to Nichol to say that "there can be little if any doubt as to the resolvability of the nebula. Since you left us, there was not a single night when, in absence of the Moon, the air was fine enough to admit of our using more than half the magnifying power the speculum bears: still, we could plainly see that all about the trapezium is a mass of stars; the rest of the nebula also abounding with stars, and exhibiting the characteristics of resolvability strongly marked. Rosse".[28]

The alleged resolution of the Orion Nebula, confirmed soon afterwards at Harvard, marked a turning point. Some astronomers, like Nichol, argued that this meant the end of the nebular theory. "Every shred of that evidence", writes

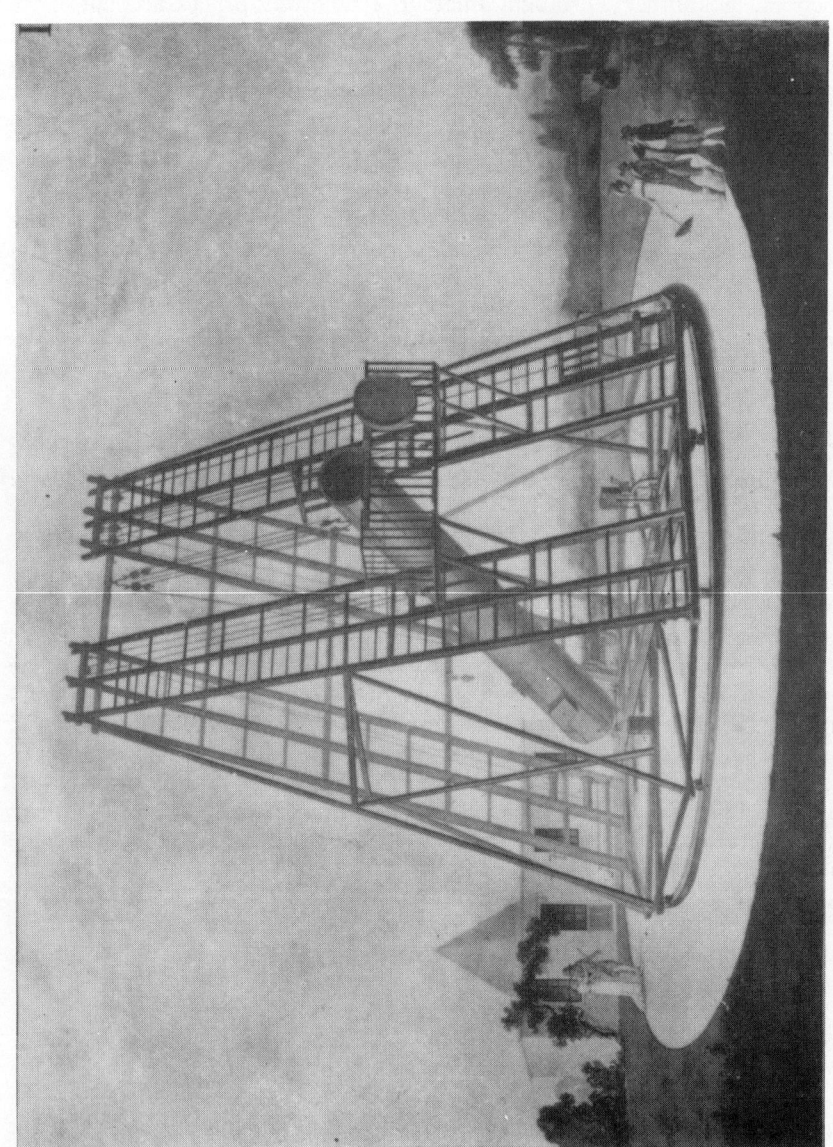

FIG. 6. William Herschel's 25ft reflector made for the King of Spain (from the watercolour accompanying the instructions for use).

FIG. 7. A contemporary photograph of the Birr reflector with 6ft mirror, "the Leviathan of Parsonstown" (Royal Astronomical Society archives).

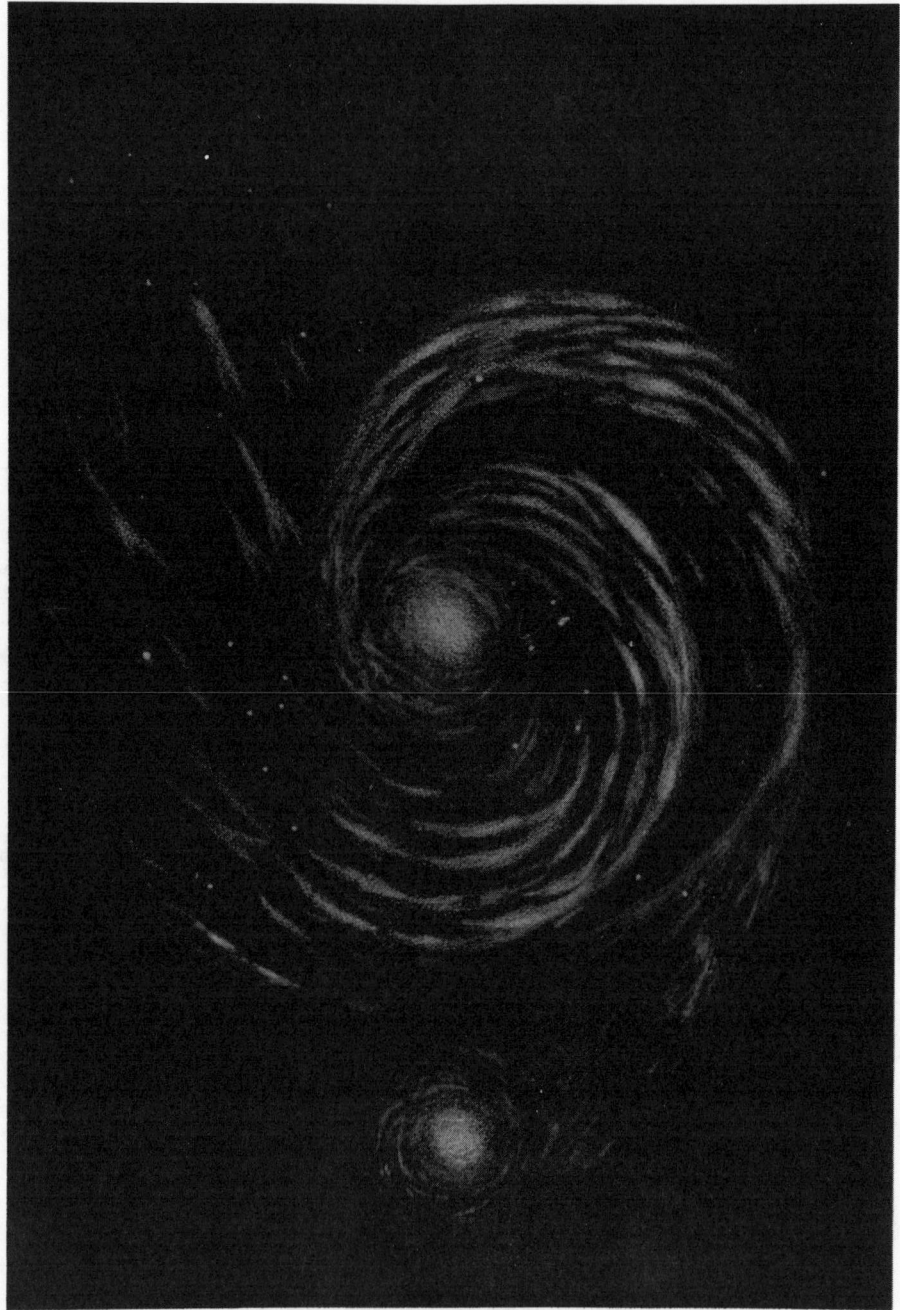

FIG. 8. A Birr drawing of M51. This was the first of the spiral nebulae to be recognised as such (in 1845, with the newly-completed 6ft).

Nichol, "which induced us to accept as a reality, accumulation in the heavens of matter *not stellar,* is for ever and hopelessly destroyed".[29] John Herschel, in his immensely influential *Outlines of astronomy* (1849), admitted that "it may very reasonably be doubted whether there be really any essential physical distinction between nebulae and clusters of stars".[30] Even those who continued to insist that isolated observations were incapable of disproving the nebular theory, mostly admitted that the existence of island universes had been demonstrated by these observations. Looking back half a century later, Charles Young of Princeton commented that "In some respects this old belief strikes one as grander than the truth even. It made our vision penetrate more deeply into space than we now dare think it can".[31]

Yet there were dissenting voices. For John Herschel, the Magellanic Clouds to which he had devoted so much attention while at the Cape of Good Hope were a stumbling block to the acceptance of Robinson's claims, for the Magellanics contained a mixture of stars and (so it seemed) true nebulosity. In the *Outlines,* after presenting in full detail the supposed observations of Rosse and supporting as unequivocally as he ever expressed any view the generalizations drawn from them, he writes of the Magellanics: it is a "demonstrated fact, that stars of the 7th or 8th magnitude and irresolvable nebula [*sic*] may co-exist within limits of distance not differing in proportion more than as 9 to 10, a conclusion which must inspire some degree of caution in admitting, *as certain,* many of the consequences which have been rather strongly dwelt upon in the foregoing pages".[32]

Other doubts arose from changes which – once more – were thought to have been observed in the Orion Nebula. As John Herschel had pointed out in 1826, a

FIG. 9. The Crab Nebula, M1, as seen in the Birr 3ft mirror (*Philosophical transactions* for 1844, Plate XVIII): "It is studied with stars, mixed however with a nebulosity probably consisting of stars too minute to be recognized" (p. 322).

Fig. 10. The Owl Nebula, M97, as seen in the Birr 6ft mirror (*Philosophical transactions* for 1850, Plate XXXVII).

nebula can look very different when viewed with different telescopes on the same night, or with the same telescope on different nights;[33] while drawings or engravings could never give a completely faithful representation of what had been observed. And throughout his career he continued, rightly, a sceptic of the changes alleged in the Orion Nebula. Forceful illustrations of how different the nebulae can appear in reflectors of different power were contained both in Rosse's discovery of the spiral shape of some nebulae (Figure 8), and in the bizarre sketches made at Birr of nebulae like the Crab and the Owl (see Figures 9 and 10). Nevertheless, at Pulkova Otto Struve was convinced that his colleague M. Liapounov had observed genuine changes in the Orion Nebula, in which case the alleged resolution of the nebula into stars by the Birr telescope must be mistaken. In the winter of 1850/51 Struve wrote two letters to Rosse reporting Liapounov's belief "that the nebula has undergone several remarkable changes since the drawing of Sir John [Herschel]". To his credit, Rosse carefully preserved the letters by pasting them in his Astronomical Diary.[34] In 1853 Struve made his position public:

> Recently astronomers have believed they have found, in the resolution into stars of a small number of nebulae with the help of fine modern telescopes, a complete refutation of the ideas of Sir William Herschel; although to some extent the alleged miracles of resolution, for example of the Orion and Andromeda Nebulae, are nothing but illusions. The study of a small number of nebulae will by no means settle the question of the nature of these objects. To do this we must examine and compare the 6000 nebulae now known, or at least all those visible from a given place. Meanwhile, we believe that the discoveries of M. Liapounov have given, to the contrary, a direct proof of the truth of William Herschel's view. For since in certain nebulae there are major rapid changes, these nebulae are not clusters of stars; for the starry heavens, even in the regions nearest to us, present only changes of position that are extremely slow and comparatively small.[35]

But the Birr astronomers were as justified in their scepticism of Liapounov's alleged changes as Struve was in his doubts of the Rosse resolution: both sides believed they were basing their opinions on hard observational fact, but both were deceived. Nor did Struve fare any better with the fourth Earl, to whom he suggested in 1869 that instead of speaking of the resolvability into stars of the Orion Nebula, one should say "there is a tendency of the nebulous matter to form itself in separate knots sometimes in this, sometimes in an other direction".[36] Wise words, and indeed (as we shall see) part of the significance of Hubble's discovery of a Cepheid variable star in the Andromeda Nebula in the winter of 1923/24 was that the star-like object had the very familiar and characteristic light curve of a Cepheid variable, and therefore was a true star and not just a knot of nebulosity.

Whatever the truth about the alleged changes in the Orion Nebula, the astronomical world was impressed and shaken a few years later by an altogether more convincing example of a variable nebula. In October 1852 J. R. Hind had reported the discovery of a small nebula in Taurus. This nebula, NGC 1555, was observed several times in the following years, and especially by H. L. D'Arrest. Imagine, then, D'Arrest's astonishment in October 1861 when he could find no trace of it! News of the disappearance spread quickly, and at the end of the year Otto Struve found he could just make out a faint trace of the nebula; by the following March he considered it had grown brighter.[37]

"Hind's wonderful nebula",[38] as one contemporary called it, provided an example of change among the nebulae of unimpeachable authenticity. Other examples were now taken more seriously, and by the end of 1863 several accounts of island universes had appeared in which a new hesitation was evident. "A few years since", we read in the *American annual cyclopaedia*, "the resolution, by aid of Lord Rosse's powerful telescope, of one or more of those singular, fixed, and hazily luminous patches in remote space into congeries of actually separated and luminous stars, a resolution confirmed subsequently by other instruments and on other nebulae, proved sufficient to shake the nebular hypothesis of Laplace to its foundation. ... Now, however, assuming that the three nebulae referred to [Hind's and two others] have actually disappeared or faded, new and strange questions are raised; and the facts seem likely to prove as irreconcilable with the doctrine that all nebulae are clusters, as the resolution of some of them was with the hypothesis of Laplace".[39]

The next year the 'new astronomy' of astrophysics was to demonstrate at a stroke the existence of 'true nebulosity'. William Huggins, the latest of the amateurs to make distinguished contributions to the 'riddle of the nebulae', applied a spectroscope to the light of a nebula for the first time:

> On the evening of the 29th of August, 1864, I directed the telescope for the first time to a planetary nebula in Draco. The reader may now be able to picture to himself to some extent the feeling of excited suspense, mingled with a degree of awe, with which, after a few moments of hesitation, I put my eye to the spectroscope. Was I not about to look into a secret place of creation?
>
> I looked into the spectroscope. No spectrum such as I expected! A single bright line only! At first, I suspected some displacement of the prism, and that I was looking at a reflection of the illuminated slit from one of its faces. This thought was scarcely more than momentary; then the true interpretation flashed upon me. The light of the nebula was monochromatic, and so, unlike any other light I had as yet subjected to prismatic examination, could not be extended out to form a complete spectrum. After passing through the two prisms it remained concentrated into a single bright line, having a width corresponding to the width of the slit, and occupying in the instrument a position at that part of the spectrum to which its light belongs in refrangibility. A little closer looking showed two other bright lines on the side towards the blue, all the three lines being

separated by intervals relatively dark.

The riddle of the nebulae was solved. The answer, which had come to us in the light itself, read: Not an aggregation of stars, but a luminous gas.[40]

Acknowledgements

I am grateful to the Royal Astronomical Society for permission to use material from the Herschel archives, and to the Earl of Rosse for every assistance with the archives at Birr Castle.

REFERENCES

1. F. G. W. Struve, *Études d'astronomie stellaire* (St Petersburg, 1847), 48.
2. For an account of this, see Simon Schaffer, "Uranus and the establishment of Herschel's astronomy", *Journal for the history of astronomy*, xii (1981), 11-26.
3. Though he was persuaded to omit most of the speculations from the published paper. See J. L. E. Dreyer (ed.), *The scientific papers of Sir William Herschel* (London, 1912), i, pp. xc-xci.
4. See chap. 1 above, ref. 10.
5. *Mémoires de l'Académie royale des Sciences* for 1784, 333.
6. See the article by J. A. Bennett cited in chap. 1 above, ref. 4.
7. F. T. Schubert, *Traité d'astronomie theorique* (French trans., St Petersburg, 1822), ii, 37.
8. On this, see Simon Schaffer, "Herschel in Bedlam: Natural history and stellar astronomy", *British journal for the history of science*, xiii (1980), 211-39.
9. Letter of Watson to Herschel, 7 September 1789 (Royal Astronomical Society Herschel MSS, 13.W.57).
10. W. Herschel, "On the construction of the heavens", *Philosophical transactions*, lxxv (1785), 213-66, p. 260.
11. Herschel, *ibid.*, 216-17. On repulsive forces, see his "Catalogue of a second thousand of new nebulae and clusters of stars; with a few introductory remarks on the construction of the heavens", *Philosophical transactions*, lxxix (1789), 212-55, pp. 221-2, and Dreyer (ed.), *op. cit.*, i, p. lxxv. *Cf.* Simon Schaffer, "'The great laboratories of the universe': William Herschel on matter theory and planetary life", *Journal for the history of astronomy*, xi (1980), 81-111.
12. W. Herschel, "On nebulous stars, properly so called", *Philosophical transactions*, lxxxi (1791), 71-88. The 'nebulous star' was a planetary nebula with a prominent central star and a halo of no less than 3' diameter. It therefore appeared significantly different from the planetary nebulae already known to him, which had diameters measured in seconds and the appearance of a faint disk.
13. W. Herschel, "Astronomical observations relating to the construction of the heavens, arranged for the purpose of a critical examination, the result of which appears to throw some new light upon the organization of the celestial bodies", *Philosophical transactions*, ci (1811), 269-336, and "Astronomical observations relating to the sidereal part of the heavens, and its connection with the nebulous part: arranged for the purpose of a critical examination", *ibid.*, civ (1814), 248-84. Herschel uses the plant analogy in the final paragraph of his 1789 paper (ref. 11): "For ... it is not almost the same thing, whether we live successively to witness the germination, blooming, foliage, fecundity, fading, withering, and corruption of a plant, or whether a vast number of specimens, selected from every stage through which the plant passes in the course of its existence, be brought at once to our view?" In introducing his 1811 paper, he says: "...it will be found that those [celestial objects] contained in one article, are so closely allied to those in the next, that there is perhaps not so much difference between them, if I may use the comparison, as there would be in an annual description of the human figure, were it given from the birth of a child till he comes to be a man in his prime" (p. 271).
14. J. F. W. Herschel, "Descriptions and approximate places of 321 new double and triple stars", *Memoirs of the [Royal] Astronomical Society of London*, ii (1826), 459-97, pp. 487-8.
15. "The quantity of emitted particles [of light] may well become adequate to the constitution of a shining fluid, or luminous matter, provided a cause can be found that may retain them from flying off, or reunite them" (Herschel, "On nebulous stars", 87). On this see Schaffer, *op. cit.* (ref. 11), 90-93.
16. See Schaffer, *ibid.*, 93-96.
17. *Memoirs of the [Royal] Astronomical Society of London*, i (1822-25), 4. Herschel was remarkably frank about his intention to speculate: "...if we add observation to observation,

without attempting to draw not only certain conclusions, but also conjectural views from them, we offend against the very end for which only observations ought to be made. I will endeavour to keep a proper medium; but if I should deviate from that, I could wish not to fall into the latter error" (*op. cit.* (ref. 10), 213). He similarly repudiated the requirement to keep observations uncontaminated by speculation: "It may even be said, that since observations are made with no other view than to draw such conclusions from them as may instruct us in the nature of the things we see, there cannot be a more proper time for entertaining surmises than when the object itself is in view" ("Observations tending to investigate the nature of the Sun", *Philosophical transactions*, xci (1801), 265-318, p. 269).

18. For a biography of John Herschel, see Gunther Buttmann, *The shadow of the telescope* (New York, 1970).
19. G. B. Airy, "An Address ... on presenting the honorary medal to Sir J. F. W. Herschel", *Memoirs of the Royal Astronomical Society*, ix (1836), 303-12, pp. 304-5.
20. On the astronomy of the third and fourth Earls of Rosse, see Patrick Moore, *The astronomy of Birr Castle* (London, 1971). Much material relating to the telescopes is in the published papers of the third Earl, to be found in *The scientific papers of William Parsons, third Earl of Rosse 1800-1867* (London, 1926). The surviving unpublished papers and instruments have recently been catalogued (to appear in *Journal for the history of astronomy*) by Dr J. A. Bennett and the present writer.
21. See the article by J. A. Bennett cited in chap. 1 above, ref. 4.
22. Earl of Rosse, "Observations on some of the nebulae", *Philosophical transactions* for 1844, 321-4, p. 324.
23. From the manuscript entry in *Astronomical scrapbook: 3rd and 4th Earl of Rosse*, preserved at Birr Castle.
24. *Ibid.*
25. T. R. Robinson, "On Lord Rosse's telescope", *Proceedings of the Royal Irish Academy*, iii (1845), 114-33, p. 130.
26. T. R. Robinson, ["On Lord Rosse's telescope"], *ibid.*, iv (1848), 119-28, p. 119.
27. J. F. W. Herschel, *A treatise on astronomy* (London, 1833), para. 619.
28. J. P. Nichol, *Architecture of the heavens* (10th edn, London, 1850), 112-14.
29. *Ibid.*, 116.
30. J. F. W. Herschel, *Outlines of astronomy* (London, 1849), para. 871.
31. Charles A. Young, *A text-book of general astronomy* (2nd edn, Boston, 1898), 560.
32. Herschel, *Outlines*, para. 894.
33. Herschel, *op. cit.* (ref. 14), 488-9.
34. Preserved at Birr Castle.
35. *Mélanges mathématiques et astronomiques tirés du Bulletin de l'Académie impériale des Sciences de St. Pétersbourg*, ii (1853), 47-48. On the other hand, in the 122-page monograph he later wrote with Liapounov ("Observations de la Grande Nébuleuse d'Orion", *Mémoires de l'Académie impériale des Sciences de St. Pétersbourg*, sér. vii, vol. v, no. 4 (1862)), Struve displays exemplary caution: "L'extrait précédent de mon journal d'observations contient sans doute des indications très fortes de changements dans l'état de la nébuleuse. Néanmoins je suis bien loin de prétendre que tous les changements notés soient élevés au dessus de tout doute.... Malgré la bonne volonté de se tenir libre de tout préoccupation, l'imagination, supportée dans ces cas par l'insuffisance de nos moyens d'observation et par l'effet de l'état variable de l'atmosphère, nous entraîne facilement à voir ce que nous voulons voir ou plutôt a ce qui s'accorde le mieux avec nos pensées intimes et d'un autre côté à négliger de noter ce qui paraît s'opposer à nos vues" (pp. 111-12).
36. The letter is pasted into the *Astronomical scrapbook* (ref. 23).
37. See for example Herschel's *Outlines* (10th edn, London, 1869), app. Note K.
38. T. W. Webb, writing in *The intellectual observer*, iv (1863-64), 56.
39. *The American annual cyclopaedia and register of important events of the year 1862* (New York, 1863), 175-6.
40. William Huggins, *The scientific papers of Sir William Huggins* (London, 1909), 106, reprinted from *The nineteenth century review* for June 1897.

3. Island Universes: An Overview

Spectroscopy had shown in 1864 that some nebulae are truly gaseous. But are others, nevertheless, star systems outside of and independent of our Galaxy? – perhaps even comparable systems, 'island universes', that in William Herschel's phrase "may well outvie our milky-way in grandeur".[1]

Herschel had answered in the affirmative in the late 1780s because he had a theory of our Galaxy and a theory of nebulae and so could compare the two. His telescopes, he assumed, could reach to the borders of our Galaxy which was therefore of limited extent. The visible nebulae were believed to be star systems, and those that appeared milky were very distant; therefore milky nebulae, like those of Orion and Andromeda, which even though distant were spread across a large area of the sky, must be star systems of enormous extent and may well "outvie" our Galaxy.

All this changed after 1790. His recently-completed 40ft telescope had brought many more stars of the Milky Way into view, and he had therefore been wrong to suppose his 20ft had reached the borders in every direction.[2] Indeed, even in the 40ft the Galaxy was in some directions "fathomless".[3] It was, therefore, "the most brilliant, and beyond comparison the most extensive sidereal system",[4] an immense stratum of stars, with "two comparatively vacant spaces on each side, the situation of globular clusters, of planetary nebulae [here assumed to be star clusters], and of far extended nebulosities".[5] In the visible universe, therefore, there could be no comparable galaxies.

Could there nevertheless be among the nebulae, distant star systems outside of and independent of the Galaxy? And if so, how could they be distinguished from the true (and nearby) nebulae he now knew to exist? Herschel might recognise the Orion Nebula as a small, nearby nebulosity because of its changes in shape;[6] but a very distant star system would be indistinguishable from unchanging nebulosity.[7]

True; but this notwithstanding, Herschel was satisfied that distant star systems did exist, and indeed in his last papers the trend is towards more rather than fewer star systems: "It seems to be highly probable that some of the cometic, many of the planetary, and a considerable number of the stellar nebulae, are clusters of stars in disguise."[8] We may suspect that infinite reaches of empty space made nonsense; and unless the Galaxy was a 'black hole' and by its attractive force held back the starlight from escaping,[9] light must travel out to fill the universe and eventually condense into distant stars.[10] In 1799 he had calculated that his 40ft could reach a hypothetical cluster of 50,000 stars so far away that its light would take two million years to reach us;[11] in 1802 he drew the implication that big telescopes have "a power of penetrating into time past"[12] and that such a cluster would have to have existed two million years ago; and in 1813 he told the poet Thomas Campbell: "I have observed stars of which the light, it can be proved, must take two million years to reach the earth."[13] Herschel therefore believed that the Galaxy was unique in the observable

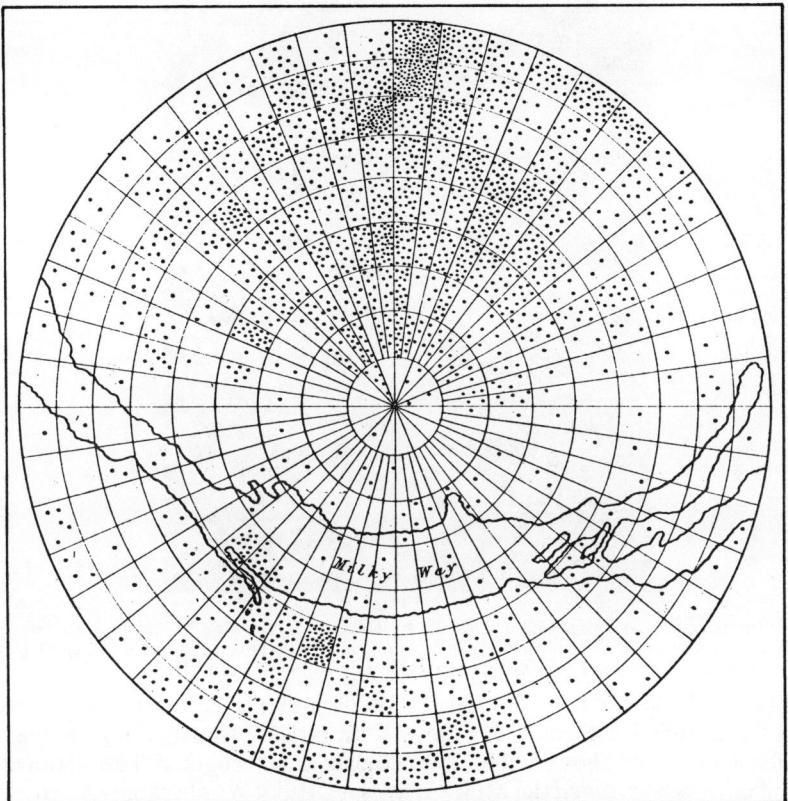

FIG. 1. Chart of the Northern Hemisphere showing the distribution of nebulae and their tendency to congregate away from the Milky Way (by R. A. Proctor, in *Monthly notices of the Royal Astronomical Society*, xxix (1869), 337-44).

universe, but that independent star systems, small by comparison, existed as far as telescopes could reach.

In the middle of the nineteenth century, the pendulum had swung once more. John Herschel, whose years at the Cape had made him the foremost living observer of the Milky Way, believed that in most directions he had seen through the Milky Way and beyond,[14] so that the Milky Way was once more regarded as of limited extent; and Lord Rosse's claims had restored the Orion Nebula and others to their former status as vast star systems, island universes in space.

In the last decades of the century, the pendulum swung back once again. Huggins had shown that *some* nebulae are gaseous, and with the scientist's predilection for a simple theory he then went on to suggest that even the Andromeda Nebula, which gave enough light to reveal a continous spectrum, was composed not of stars but of gaseous matter under special physical conditions.[15]

This reluctance to accept that although *some* nebulae are gaseous, *others* may be independent island universes received support from two observational factors. The first (see Figure 1) was a more detailed study of the distribution of

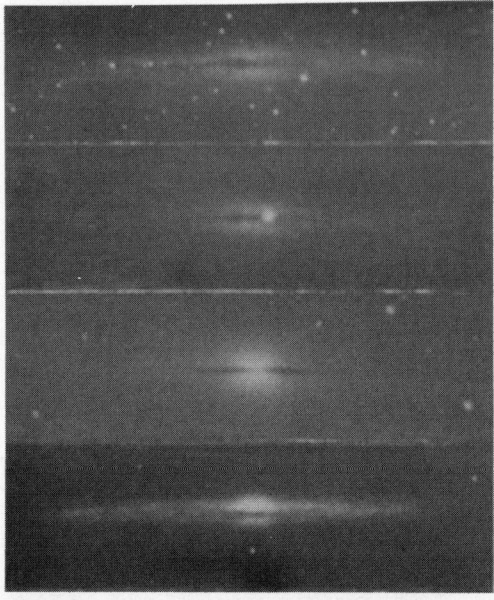

FIG. 2. Photographs of spiral nebulae taken by Curtis and showing "a band of absorbing or occulting matter". From "A study of occulting matter in the spiral nebulae", by H. D. Curtis, *Publications of the Lick Observatory*, xiii, part 2 (1918), Plate III.

nebulae in the sky, which showed, as William Herschel had noted a century earlier, that nebulae cluster near the galactic poles.[16] Investigations in the late nineteenth century showed that the nebulae which might still be claimed as 'island universes' *avoided* the Milky Way. The Milky Way became known as the "Zone of avoidance" and astronomers asked why, if these nebulae were other island universes, they should be located in a manner specially related to *our* island universe. Secondly, in 1885 a new star flared up in the Andromeda Nebula, equal by itself to perhaps one-tenth of the entire nebula.[17] It seemed impossible that one single star could equal millions of ordinary stars, which was necessarily the case if the Andromeda Nebula was an island universe of many millions of stars. In fact the new star was a supernova, but at the time there was no reason to believe in the existence of celestial fireworks on such a scale.

As a result of this converging evidence, astronomers were agreed at the turn of the century that science knew of only one island universe, our own Galaxy, although other island universes might exist beyond the reach of telescopes. However, just before his death in 1900 James E. Keeler of Lick Observatory had taken photographs that showed that spiral nebulae (of the form first recognised by the third Earl of Rosse) existed in hundreds of thousands. Although it was generally agreed that these spirals would develop into either sparse star clusters or single stars,[18] they offered a fascinating new field of study; several investigators aimed spectrographs at the brighter spirals, and around 1912 it was conclusively demonstrated that their spectra were dominated by features well known as characteristic of stars.

In 1913, to the astonishment of all astronomers and the incredulity of some, V.

M. Slipher of Lowell Observatory discovered from the shifts of the spectral lines of the Andromeda Nebula that the nebula was rushing towards the observer at 300 km per second, the highest speed then known for any astronomical body. Slipher obtained similar values for other spirals, and these radial (line-of-sight) velocities were on average so much larger than those of the confirmed members of the Galaxy, that many astronomers concluded the spirals could not be part of the Galaxy's dynamical system and, in view of their spectra, must be island universes.

In 1917 the island universe theory received renewed support when Heber D. Curtis of Lick Observatory and G. W. Ritchey of Mount Wilson discovered faint novae in spiral nebulae (the episode is discussed in detail in the following chapter). Old plates were now eagerly scrutinised for other novae in spirals, to such effect that Curtis satisfied himself that the 1885 nova in the Andromeda Nebula was abnormal, and he calculated from the much smaller brightness of the other novae that the spirals lay far beyond the Galaxy.

Curtis had also pursued a programme of nebular photography. In photographs of nebulae seen edge-on (see Figure 2) he had repeatedly found dark lines indicating obscuring matter, and he realised that a similar ring of obscuring matter in the plane of our own Galaxy could explain the zone of avoidance: we fail to see other island universes in the direction of the Milky Way, not because for some curious reason the island universes do not exist in such directions, but because we cannot see them through the obscuration.

Curtis had now answered the two items of evidence seen in the last years of the nineteenth century as decisive proof that the spirals are not island universes. Indeed, by the end of 1917, most of the leading students of spiral nebulae agreed with Curtis. To them the combined testimony of spectra, radial velocities, novae, and obscuration, was persuasive and clear.

As we have seen, changes in nebulae had been mistakenly "observed" since the time of William Herschel; but now the application of photography seemed to provide objective evidence of motions in spiral nebulae. In 1916, Adriaan van Maanen of Mount Wilson Observatory had used a stereocomparator which in effect superimposed two photographs of the same spiral, M101.[19] Van Maanen concluded (see Figure 3) that the nebula exhibited internal motions that could be interpreted either as outward motions along its spiral arms or as the result of rotation. At first this finding was not widely regarded as detrimental to the island universe theory, and at the time of the so-called Great Debate on "The Scale of the Universe" between Curtis and Harlow Shapley in 1920 (see chap. 5 below), the spiral nebulae were usually viewed as external galaxies. But van Maanen's contribution became a leading issue in the early 1920s when he measured several more spirals (see Figure 4) and found his results agreed excellently with those for M101;[20] furthermore, it was realised how well the (supposedly) observed motions agreed with an elegant theory of nebular evolution advanced by the English mathematician J. H. Jeans.

Meanwhile Shapley in 1918 had proposed a dramatic new theory of our Galaxy. Instead of, in effect, working his way outwards from the Sun via the nearer stars towards the more distant as had his immediate predecessors, Shapley proposed that the globular clusters congregate in one half of the sky because they surround the centre of the Galaxy and that centre is therefore far

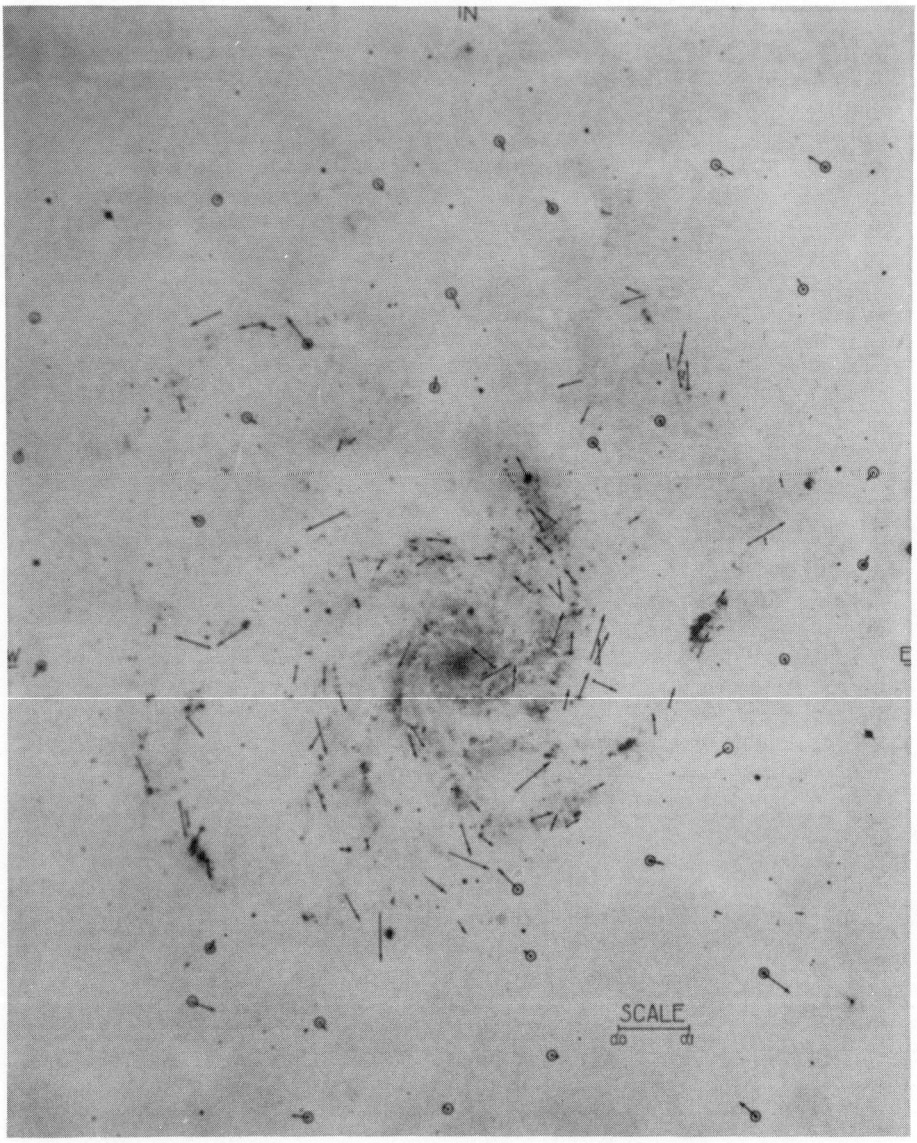

FIG. 3. Movements in the spiral nebula M101, derived in 1916 by A. van Maanen by using a stereocomparator to (in effect) superimpose similar photographs of different date.[19]

away in the general direction of the globular clusters.[21] By this bold stroke Shapley in effect reached out above and below the plane of the Galaxy, to the globular clusters which lay in directions where it seemed there was little obscuration to obstruct the view. Shapley used various techniques to determine the distances of the clusters, notably the new-found tool of Cepheid variable stars which were believed by many astronomers to exhibit a relationship

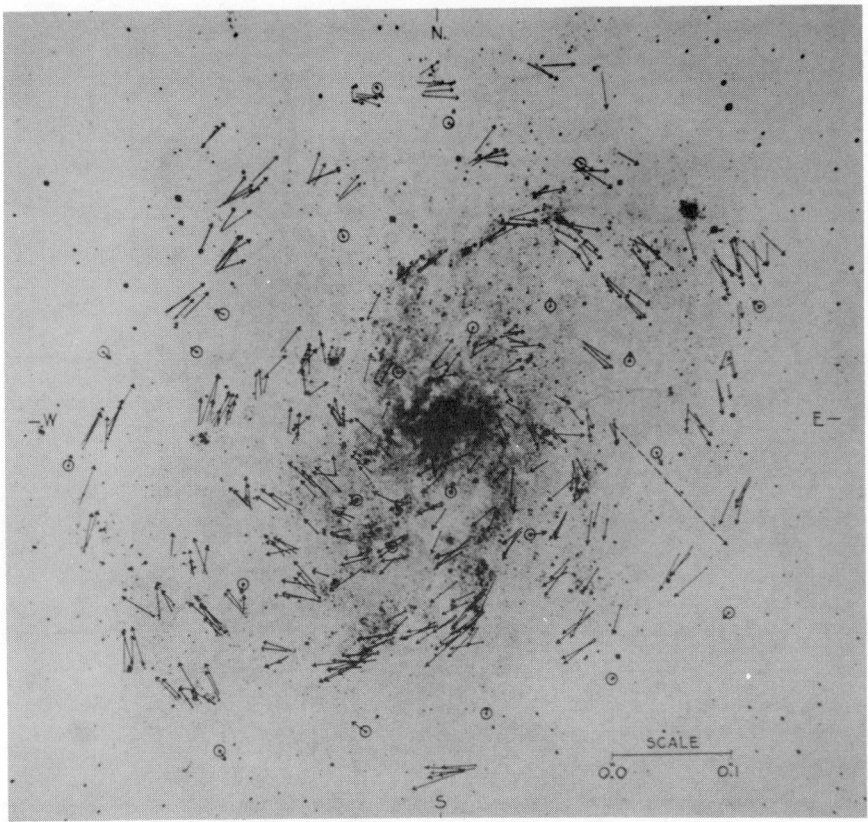

Fig. 4. Movements in M33, similarly obtained by van Maanen in 1923 (*Contributions from the Mount Wilson Observatory*, no. 260).

between the period of light variation and absolute brightness: once the period of a Cepheid was observed, its absolute brightness was known, and its apparent brightness was a good indication of its distance. He arrived at a diameter for the Galaxy of some 300,000 light years (see chap. 5 below), many times greater than previous estimates which had been vitiated by the obscuration in the plane of the Galaxy which conceals the more distant stars. Years later it was found that the light from Shapley's clusters was also reduced by obscuration so that the clusters seemed fainter and more distant than they really were, and his diameter for the Galaxy has been cut to one-third. But meanwhile his bold stroke multiplied the accepted size of the Galaxy by an order of magnitude, and made it all the more unlikely that any spiral nebula could be a *comparable* island universe. For van Maanen's spirals to be comparable island universes, the velocities of the outlying regions would have to be greater than the speed of light. For some astronomers this was plainly absurd, and van Maanen's rotations became a major argument against island universes.

Although it is hard to imagine scientific results closer to the plain facts of

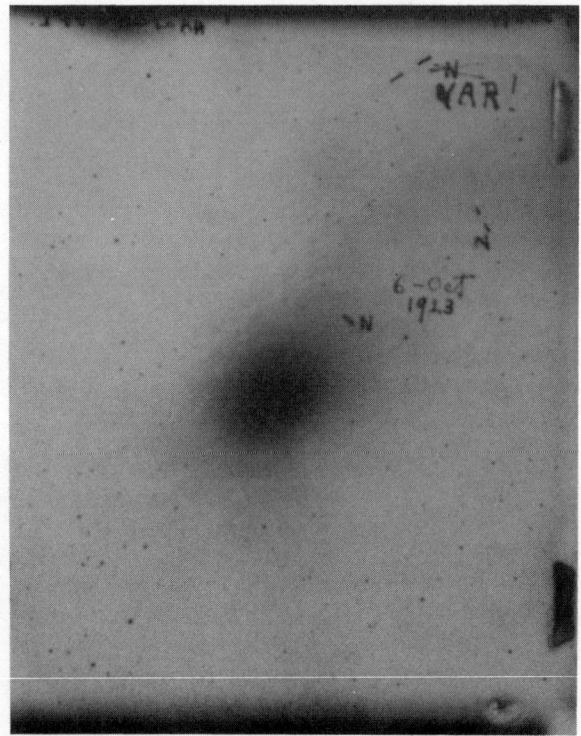

FIG. 5. The photograph of the Andromeda Nebula, taken in October 1923, on which Hubble discovered the first Cepheid variable star in a spiral nebula. Hubble has marked the star "N" (for nova) but has later altered this to "VAR!".

observation, other astronomers regarded the internal motions with strong reservations. They were almost the only substantial evidence against the spirals as island universes; and how else could one interpret the novae, radial velocities, and spectrum observations? So, while the island universe theory lost some support in the early 1920s, many astronomers judged it to be on balance the best available explanation of the spirals, especially as researches favourable to the theory continued to be forthcoming.

The sudden, and totally unexpected, resolution of the debate on island universes was brought about in 1923 and 1924 by a series of startling discoveries made by Edwin Hubble at Mount Wilson with the 60" and 100" reflectors. In October 1923 a systematic attack on novae in the Andromeda Nebula was launched with the 100", and on the first good plate (see Figure 5) Hubble discovered two ordinary novae and a third, faint star that proved on investigation to be a true variable. Hubble's worksheets (Figure 6) show that the variable had been present, unrecognised, on nearly seventy earlier Mount Wilson plates going back to the early days of the 60" in 1909. By the end of October he had established its approximate period (one month); and he must have recognised that its light curve resembled the familiar light curve of a

Cepheid variable, and so was beyond question a true star and no mere knot of nebulosity. Furthermore, the theory of Cepheid variables that Shapley had used to such effect in measuring great distances decreed that Hubble's star, with its long period, was intrinsically very bright despite its faint appearance. Its distance, therefore, and the distance of the nebula of which it was part, must be enormous, around one million light years; and since the nebula extends so far across the sky despite its enormous distance, it must be another island universe.

For the present, however, Hubble kept his dramatic discovery to himself, no doubt anxious himself to observe the light of the variable perform the upward leap characteristic of a Cepheid. This he did by means of photographs taken every night between 2 and 7 February (see Figure 6). He was now able (see Figure 7) to establish the light curve and to derive a very accurate value for the period. The resulting distance he gives as 330,000 parsecs, well over one million light years.

On 19 February, Hubble broke silence. As it was to be nearly a year before he ventured into print, his letter to Shapley (a copy is in the Hubble Archives) is a valuable expression of the state of his investigations:

> You will be interested to hear that I have found a Cepheid variable in the Andromeda Nebula (M31). I have followed the nebula this season as closely as the weather permitted and in the last five months have netted nine novae and two variables.... The two variables were found last week [a white lie!]. No. 1 is roughly 16′ preceeding the nucleus, but well within the borders of the arms, and is situated on a background of faint mottled nebulosity. Magnitudes were estimated rather hastily from a set of comparison stars and a light curve was constructed covering all available observations from 1909 down to date. The zero point in the comparison scale was extrapolated from your pg magnitudes, making what seemed to be a fair allowance for distance correction. I believe the range [?] of the variable cannot be as much as 0.3 in error, nor the median magnitude more than 0.5.
>
> Enclosed is a copy of the normal light curve, which, rough as it is, shows the Cepheid characteristics in an unmistakable fashion. Am I right in supposing it a typical cluster type curve? The period of 31.415 days corresponds to $M = -5.0$ on your period-luminosity curve. The medium sg magnitude of about 18.5 needs an uncertain correction for color index. Seares suggests 0.9 as a maximum, although your period-color curve for the Magellanic cloud calls for a higher value. With Seares value, the median pv magnitude is 17.6 and the distance comes out something over 300,000 parsecs. If the stars were dimmed materially by shining through nebulosity the distance would be correspondingly reduced....

During the remainder of 1924, knowledge of Hubble's results spread among what would today be termed the 'invisible college' of his colleagues and associates, but it was only at the urging of friends (including Shapley) that he allowed a paper to be read for him *in absentia* at the American Astronomical Society on 1 January 1925.[22] "The real reason for my reluctance in hurrying to press", he told H. N. Russell, "was, as you may have guessed, the flat contradiction to van Maanen's rotations. The problem of reconciling the two sets of data has a certain fascination, but in spite of this I believe that the

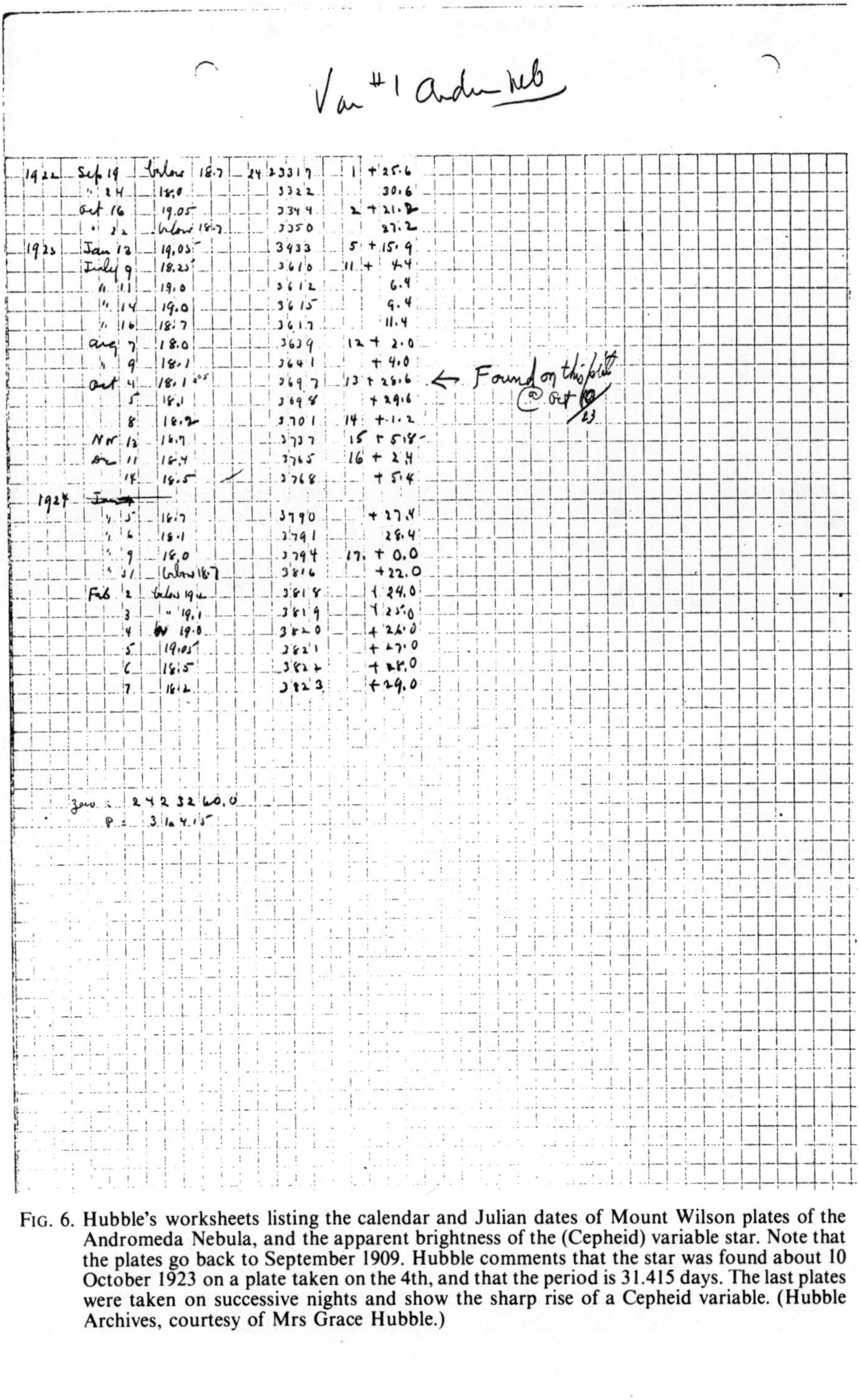

Var #1 Andr Neb

1922	Sep 19	below 18.7	2423317	1	+ 25.6
	" 24	18.0	3322		30.6
	Oct 16	19.05	3344	2	+ 21.2
	" 22	below 18.7	3350		27.2
1923	Jan 12	19.05	3433	5	+ 15.9
	July 9	18.25	3610	11	+ 4.4
	" 11	19.0	3612		6.4
	" 14	19.0	3615		9.4
	" 16	18.7	3617		11.4
	Aug 7	18.0	3639	12	+ 2.0
	" 9	18.1	3641		+ 4.0
	Oct 4	18.1 0.05	3697	13	+ 28.6 ← Found on this plate @ Oct 10/23
	5	18.1	3698		+ 29.6
	8	18.2	3701	14	+ 1.2
	Nov 13	18.7	3737	15	+ 5.18
	Dec 11	18.4	3765	16	+ 2.4
	14	18.5	3768		+ 5.4
1924	~~Jan~~				
	" 5	18.7	3790		+ 27.4
	" 6	18.1	3791		28.4
	" 9	18.0	3794	17.	+ 0.0
	" 31	below 18.7	3816		+ 22.0
	Feb 2	below 19.2	3818		+ 24.0
	3	" 19.1	3819		+ 25.0
	4	19.0	3820		+ 26.0
	5	19.05	3821		+ 27.0
	6	18.5	3822		+ 28.0
	7	18.2	3823		+ 29.0

Zero = 2423260.0
P = 31.415

FIG. 6. Hubble's worksheets listing the calendar and Julian dates of Mount Wilson plates of the Andromeda Nebula, and the apparent brightness of the (Cepheid) variable star. Note that the plates go back to September 1909. Hubble comments that the star was found about 10 October 1923 on a plate taken on the 4th, and that the period is 31.415 days. The last plates were taken on successive nights and show the sharp rise of a Cepheid variable. (Hubble Archives, courtesy of Mrs Grace Hubble.)

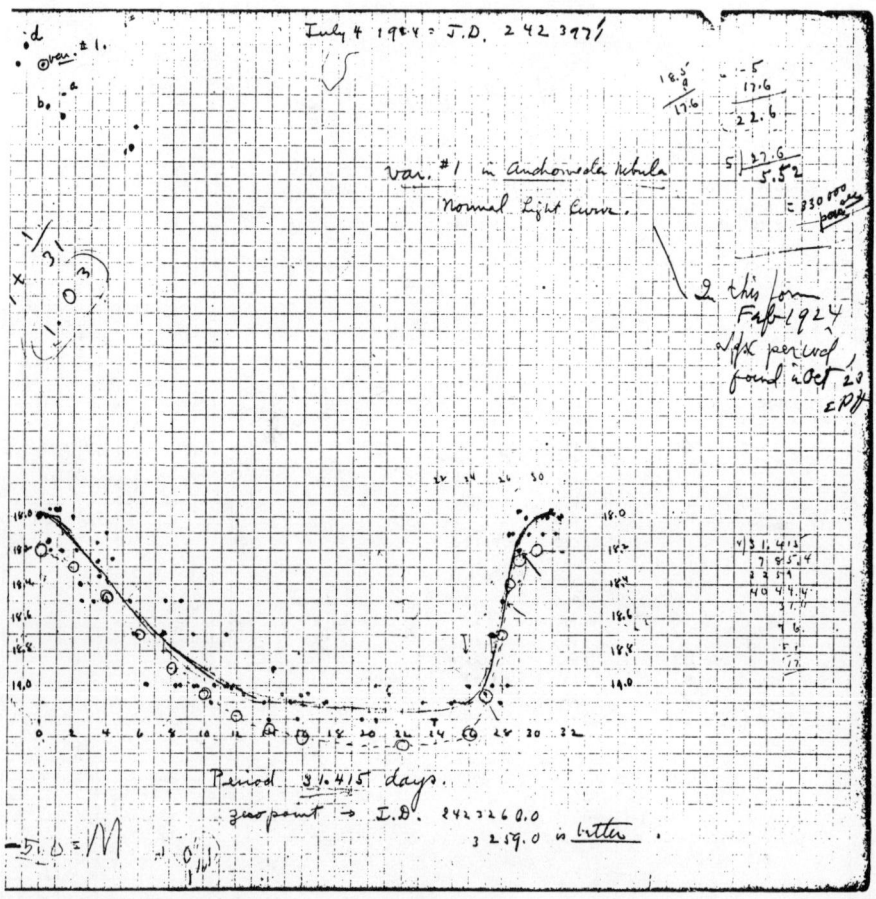

FIG. 7. Hubble's sketch of the light curve of the Cepheid variable in the Andromeda Nebula, based on the data in Figure 6. Note the calculation of a distance of 330,000 parsecs (about 1,076,000 light years), and the comment: "In this form Feb 1924/appx period found in Oct 23/EPH" (Hubble Archives, courtesy of Mrs Grace Hubble).

measured rotations must be abandoned".[23]

Yet the van Maanen rotations remained as an anomaly demanding explanation, and in the early 1930s Hubble noted with concern "a growing tendency to consider the possibility of constructing theories to reconcile the apparent angular rotations with the currently accepted large distances" for the nebulae in question.[24] Hubble's early distrust of van Maanen's measurements had long since ripened into a deep conviction that the rotations were illusory and therefore any proposing of hypotheses to accommodate them a pernicious waste of time; and the mention in the Mount Wilson Annual Report for 1930-31 that new measures by van Maanen of the great spiral nebula Messier 101 "show a decided internal motion in the same direction as was found in his original measures of this nebula"[25] may have decided him to subject the most notable of van Maanen's results to close scrutiny. In 1932 and 1933 he retraced van Maanen's steps, reworking the same photographic plates but with important differences of technique, having new photographs specially taken which would permit the nebulae to be studied over longer intervals of time, and enlisting two colleagues to give independent support to his conclusions "that internal motions of the order predicted by van Maanen do not exist"[26] in the four nebulae in question.

Hubble of course had his views on how systematic errors had crept into van Maanen's work, but he could achieve his purpose more effectively by indirect assault, by a methodological shift of emphasis. The labours of himself and his colleagues showed that systematic errors were present *either* in their measures, *or* in van Maanen's. If the errors were theirs, then these errors must, by extraordinary coincidence, be exactly opposite to the true rotations of the nebulae, so as to mislead them into obtaining a null result; in addition, the anomaly must remain. But if the errors were van Maanen's, no such coincidence was involved, and the anomaly was removed.[27]

It was a masterly manœuvre, and Hubble prepared several informal and formal papers on the subject, the longest extending to 33 type-written pages, not counting tables.[28] But extensive public controversy between colleagues at Mount Wilson would have been unseemly, and there appeared in print only two brief notes, side by side in the *Astrophysical journal*:[29] Hubble's, briefly recording the remeasurements and delicately indicating his conclusions; van Maanen's, defensively outlining new measurements in partial confirmation of his earlier results, but conceding that "it is desirable to view the motions with reserve".[30] The last lingering difficulty against island universes had been resolved.

Acknowledgements

I am grateful for help with the above from Dr R. W. Smith, whose forthcoming study, *The expanding universe: Astronomy's 'Great Debate' 1900-1931*, will be required reading for historians of modern astronomy; and grateful also to Mrs Grace Hubble, who in 1969 gave me free access to all her husband's papers.

REFERENCES

1. W. Herschel, "On the construction of the heavens", *Philosophical transactions*, lxxv (1785), 213-66, p. 260.
2. W. Herschel, "Astronomical observations and experiments tending to investigate the local arrangement of the celestial bodies in space, and to determine the extent and condition of the Milky Way", *ibid.*, cvii (1817), 302-31, p. 326.
3. W. Herschel, "Astronomical observations and experiments...", *ibid.*, cviii (1818), 429-70, p. 463.
4. *Ibid.*, 453.
5. Herschel, *op. cit.* (ref. 2), 330.
6. W. Herschel, "Catalogue ... with remarks on the construction of the heavens", *Philosophical transactions*, xcii (1802), 477-528, p. 499.
7. Herschel, *op. cit.* (ref. 3), art. IV: "Of ambiguous celestial objects".

8. Herschel, *op. cit.* (ref. 3), 466.
9. W. Herschel, "On nebulous stars, properly so called", *Philosophical transactions*, lxxxi (1791), 71-88, p. 88: "Not to mention the great counteraction of the united attractive force of whole sidereal systems, which must be continually exerting their power upon the particles [of light] while they are endeavouring to fly off". On 'black holes' in the thinking of John Michell and William Herschel, see Simon Schaffer, "John Michell and black holes", *Journal for the history of astronomy*, x (1979), 42-43. These anticipations of modern 'black holes' antedated the more familiar work of Laplace.
10. As outlined in the preceding chapter.
11. W. Herschel, "On the power of penetrating into space by telescopes", *Philosophical transactions*, xc (1800), 49-85, pp. 83-84.
12. Herschel, *op. cit.* (ref. 6), art. VII.
13. As quoted in a letter of Campbell dated 15 September 1813, cited in Constance A. Lubbock, *The Herschel chronicle* (Cambridge, 1933), 336.
14. E.g. J. W. F. Herschel, *Outlines of astronomy* (2nd edn, London, 1849), para. 798.
15. In a paper to the Royal Astronomical Society (*Monthly notices of the Royal Astronomical Society*, xxv (1865), 155-7): "It may, therefore, be that nebulae which have little indication of resolvability, and yet give a *continuous* spectrum, such as the Great Nebula in *Andromeda*, are not clusters of suns, but gaseous nebulae which, by gradual loss of heat, or the influence of other forces, have become crowded with more condensed and opaque portions" (pp. 156-7).
16. R. A. Proctor, "Distribution of the nebulae", *Monthly notices of the Royal Astronomical Society*, xxix (1869), 337-44. Herbert Spencer had also commented on this fact in his essay, "The nebular hypothesis". On Herschel, see M. A. Hoskin, *William Herschel and the construction of the heavens* (London, 1963), 65, 75, 81-82, and Plate 2.
17. See for example the account in Agnes M. Clerke, *The system of the stars* (London, 1890), 104-6.
18. "The stage of evolution which the nebula in Andromeda represents is no longer a matter of hypothesis. The splendid photograph recently taken by Mr Roberts of this nebula shows a planetary system at a somewhat advanced stage of evolution; already several planets have been thrown off, and the central gaseous mass has condensed to a moderate size as compared with the dimensions it must have possessed before any planets had been formed" (W. Huggins, "On the spectrum ... of the Great Nebula in Orion", *Proceedings of the Royal Society*, xlvi (1889), 40-60, p. 60).
19. A. van Maanen, "Preliminary evidence of internal motion in the spiral nebula Messier 101", *Contributions from the Mount Wilson Solar Observatory*, no. 118 (1916).
20. See R. Berendzen and R. Hart, "Adriaan van Maanen's influence on the island universe theory", *Journal for the history of astronomy*, iv (1973), 46-56, 73-98; Norriss S. Hetherington, "Adriaan van Maanen on the significance of internal motions in spiral nebulae", *ibid.*, v (1974), 52-53; and R. Berendzen *et al.*, *Man discovers the galaxies* (New York, 1976), section 3.
21. Harlow Shapley, "Globular clusters and the structure of the galactic system", *Publications of the Astronomical Society of the Pacific*, xxx (1918), 42-54. For a study of the contrast between Shapley's approach and that of his older contemporaries, see E. Robert Paul, "The death of a research programme: Kapteyn and the Dutch astronomical community", *Journal for the history of astronomy*, xii (1981), 77-94.
22. The story is told, with quotations from unpublished documents, in Richard Berendzen and Michael Hoskin, "Hubble's announcement of Cepheids in spiral nebulae", *Astronomical Society of the Pacific leaflet* no. 504 (June 1971).
23. Letter of 19 February 1925 (Russell Archives, Princeton University Library), published in full *ibid.*, 11-12.
24. Page 4 of unpublished paper on "Internal motions of spiral nebulae" (c. 1933), Hubble Archives (on deposit in the Henry E. Huntington Library, San Marino, Calif.). See my "Edwin Hubble and the existence of external galaxies", *Actes* of the 12th International Congress of the History of Science, v (Paris, 1971), 49-53. Mrs Grace Hubble (private communication of 15 March 1969) quotes Hubble as saying: "When a speaker at the R.A.S. said, 'If it were not for van Maanen's measurements, Hubble's results might be accepted', I decided to make the measurements."
25. P. 100.
26. "Internal motions of spiral nebulae", abstract.
27. On the former alternative, Hubble wrote: "There would be no reasons whatever for assuming the existence of such errors unless the existence of the rotations could be established on independent grounds and actually the great body of data now available concerning the nebulae solidly opposes that contengency [*sic*]. Furthermore, if the rotations were real the errors in the null set of measures would necessarily be of precisely the proper numerical values to balance the

rotations for each of four individual measures, and in the case of one measurer, cunningly adjusted to fit the different intervals for a given nebula" (Hubble Archives, "Internal motions of spiral nebulae: discussion", p. 29).
28. "Internal motions of spiral nebulae" (ref. 24 above). The other papers are: "Measures of M81 for internal motion", 12pp.; "Measures of M33 for rotation", 6pp.; "Angular rotations of spiral nebulae", 6pp.; and a discussion of van Maanen's use of corroborative evidence, 3pp.
29. Edwin Hubble, "Angular rotations of spiral nebulae", *Astrophysical journal*, lxxxi (1935), 334-5; Adriaan van Maanen, "Internal motions in spiral nebulae", *ibid.*, 336-7.
30. *Ibid.*, 337.

4. Ritchey, Curtis and the Discovery of Novae in Spiral Nebulae

The long controversy over the existence or non-existence of galaxies external to our Milky Way system ("island universes") was ended in 1924 by Edwin P. Hubble.[1] William Herschel, in the early years of his pioneering sweeps for nebulae,[2] and Lord Rosse and his colleagues during the heyday of "the Leviathan of Parsonstown",[3] had convinced themselves that they could detect myriads of stars in some of the nebulae and had concluded that these nebulae were island universes—only to be confounded by new evidence. Hubble proved the Andromeda Nebula was an external galaxy by establishing the distance of one particular feature of the nebula: a variable star which he first noticed as a possible nova in October 1923.[4] With his long sequence of photographs taken with the great telescopes of Mount Wilson, Hubble was able to show that the star's light-curve was of a shape familiar to astronomers. It was a Cepheid variable, and there were good reasons for believing Cepheids with the period of this particular example to be highly luminous, several thousand times brighter than the Sun. As the star appeared to be extremely faint, both it and the nebula of which it was part must lie at an enormous distance, perhaps 300,000 parsecs,[5] and to appear so vast at such a distance the Andromeda Nebula must be a true external galaxy.

The discovery of the Cepheid was the culmination of the search for distance indicators in spiral nebulae (as the possible galaxies were collectively known); that is, for novae and other stars of special type whose distance might be derived by a comparison of their apparent brightness with their estimated luminosity. The method was brought forcibly to the attention of astronomers by the dramatic flare-up in 1885 in the Andromeda Nebula, described elsewhere in these pages by Kenneth Glyn Jones.[6] The nova, known as S Andromedae, rose at least to seventh magnitude; and it was estimated that if the nebula were equivalent to our own Galaxy, the nova alone would have been equal to "nearly *fifty million* such suns as our own",[7] which at the time seemed preposterous. In 1923, nearly four decades after the event and on the very eve of Hubble's discovery, the implications of S Andromedae had not been forgotten:

> If the Andromeda Nebula were a distant galaxy comparable in size with our own, we could not put its distance at anything less than 200,000 light-years. But what order of real magnitude are we to assign to a star that appears to us as of the seventh magnitude at so remote a distance? No nova ever observed would be in the least comparable to such a transcendant brilliancy, and its total and rapid disappearance would be still more strange.[8]

The second nova to be discovered in a spiral nebula was found on 12 December 1895. On that day the redoubtable Mrs Williamina Fleming of Harvard College Observatory examined a photograph of the nebula NGC 5253 in Centaurus taken the previous July and noticed, 28″ from the nucleus of the nebula, a star which was not present in a photograph taken in June.[9] In the July photograph the nova, to be known as Z Centauri, was of magnitude 7·2, but when it was re-photographed four days after Mrs Fleming's discovery it had already faded to magnitude 10·9 and was now "very low, faint and near the sun.... An examination with a prism showed that the spectrum was monochromatic, and closely resembled that of the adjacent nebula. Although the spectrum is unlike those of the new stars in Auriga, Norma, and Carina, yet the object is like them in other respects".

Z Centauri was seen visually about this time by O. C. Wendell of the same observatory[10] and estimated at about magnitude 11, but it had been detected too late to have the impact on the astronomical community of S Andromedae. However, as the distinction between these novae in spirals and the more numerous galactic novae was slowly clarified, so the implication of the 1885 and 1895 observations for the island-universe theory of spirals was clearly recognized. Many astronomers were convinced these bright novae proved that spirals were near at hand and part of our own Galaxy. So, when in 1913 V. M. Slipher of Lowell Observatory announced that the Andromeda Nebula was travelling rapidly towards us,[11] J. C. Duncan wrote to say that he had looked up the literature on S Andromedae and "it seems to me that there is nothing in the spectrum observations to conflict with the hypothesis that the light of the star was due to friction of a dark body passing rapidly through a rare cloud such as the nebula may possibly be".[12] Slipher was of the same opinion: "The result suggests that the nebula in its swift course through space might have encountered a dark 'sun' thus giving rise to the peculiar nova that appeared near the nucleus of the nebula in 1885".[13]

It was not until 1917 that other novae were detected in spirals, novae that were consistently very much fainter, and so less obviously in conflict with the island-universe theory of spirals than the stars of 1885 and 1895. Hubble, writing in 1936, gives the received account of these happenings:

> In July, 1917, Ritchey, at the Mount Wilson Observatory, found a previously unrecorded star ($m = 14.6$) on a photograph of the spiral NGC 6946, which, from further data, including a small-scale spectrogram, was identified as a nova. Both Ritchey and Curtis, the latter at the Lick Observatory, immediately examined all duplicate plates of nebulae in the large collections at their disposal, and found several earlier cases where novae had appeared in spirals.[14]

The episode, Hubble considers, is perhaps even more significant than his own subsequent discovery of Cepheids, "for the discovery of novae on photographic plates initiated the study of stars involved in nebulae. Stars were the clews which led to distances".[15] Accordingly, inside the front cover of their *Astronomy of the twentieth century*, Struve and Zebergs give as the principal event of 1917: "G. W. Ritchey discovers nova in NGC 6946; this leads to search for novae in galaxies [*better*: 'spirals'] and a method of estimating the distances of galaxies".[16]

Ritchey, whose stormy career at Mount Wilson was soon to end in his dismissal, had been examining a plate of NGC 6946 taken a few days earlier, on 19 July, when he noticed a star which was not present on a similar plate taken on 25 June.[17] The new star, unlike Z Centauri, had been caught while it was still bright enough to be studied, and the discovery was broadcast by telegram to other observatories.[18] True, the magnitude was only 14, but this was still bright enough even for astronomers with smaller telescopes to share in the excitement of the discovery; the *Observatory*, for example, printed a photograph of *Nova Ritchey* taken by Max Wolf at Königstuhl.[19]

It must have been with mixed feelings that Heber D. Curtis of Lick Observatory read the telegram from Mount Wilson on 28 July. Curtis had continued and greatly extended the programme of nebular photography which James Keeler initiated at Lick[20] before his untimely death in 1900, and he had improved the mounting of the fine Crossley reflector.[21] In 1913 Curtis had reported that he was photographing

> the larger and brighter objects given in Dreyer's New General Catalogue.... In the first place, such a permanent record of the present forms of these nebulae will give data from which future studies may be made as to the motions and changes taking place in this large and interesting class of celestial objects.... A second consideration in such a survey lies in its value for statistical studies on the proportion of nebulae of spiral form and the law of the distribution of the spirals in the sky....[22]

He was by 1914 clearly (if not yet publicly) convinced that the spirals are "inconceivably distant, galaxies of stars or separate stellar universes so remote that an entire galaxy becomes but an unresolved haze of light".[23] After all, his photographs provided a plausible explanation of the puzzling manner in which the spirals were distributed across the sky: they are absent from the plane of our Galaxy, not because they are subordinate to our Galaxy, but because they are hidden from us by obscuring matter located in our galactic plane. Curtis already had a long list[24] of examples of edge-on spirals showing "dark lane down center" and he had evidently inferred that our Galaxy would present a similar appearance to a distant observer viewing us edge-on.

The clue to the allied observational puzzle bequeathed from the previous century—the extraordinary brightness of the two novae observed in spirals—did not come to hand until March 1917. As the facts gradually unfolded, it became clear to Curtis that the novae of 1885 and 1895 were exceptional—today we recognize them as supernovae—and that most novae in spirals look much fainter and are consistent in appearance with the distances required by the island-universe theory. What happened is preserved for us in the unamended version (see Figure 1) of a rough draft by Curtis evidently intended for use in the Report for the year ended 30 June 1917:

> In March, 1917, Mr. Curtis discovered a new star of about the fourteenth magnitude which had appeared in the spiral nebula N.G.C. 4527; this star appeared at some time shortly before March 20, 1915. A search through the plates of the Crossley collection resulted in the discovery of two more new stars in spirals; these two novae appeared in the *same* spiral, N.G.C. 4321, the first one at some time prior to March 17, 1901, and the second at some time prior to March 2, 1914; these also were of about magnitude fourteen. All these objects have since disappeared. The occurrence of

such novae in the spiral nebulae is a fact of great interest, and appears to furnish weighty evidence in favor of the theory that the spiral nebulae are, in effect, separate and very remote universes, vast congeries of stars so distant that the most powerful instruments are inadequate to resolve the individual stars.[25]

The note was evidently written in the early weeks of July. This is the period one would expect for such a memorandum; but we also have positive evidence, for Curtis amended it, not only to delete any reference to NGC 4321 but also to remind the reader that the nova in NGC 4527 is the *third* to be detected in spirals; that is, he counts the novae of 1885 and 1895 as the first and second (as was to be his invariable practice[26]), but he does not yet know of Ritchey's nova.

Curtis was not a man to rush into premature publication. As he explained that autumn, writing of NGC 4527:

In March of the present year, while comparing a Crossley negative of 1915 with one taken in 1901, a star was noticed on the 1915 image, at the edge of the inner and brighter part of this spiral, which was not on the 1901 image. It seemed advisable, before announcing it as a *nova*, to secure any evidence possible from plates taken elsewhere, in order to decide whether or not it might be a variable star instead.[27]

Between March and July he elicited the help of Harvard and Yerkes Observatories, and received two plates from Yerkes for study.[28] He was ready to publish his results when the telegram from Mount Wilson arrived, and indeed he sat down the very same day and drafted a paper for the August issue of *Publications of the Astronomical Society of the Pacific*.[29] He decided he must also include in the paper his still more remarkable discovery of *two* novae in NGC 4321, notwithstanding the incompleteness of his sets of plates and an outside possibility that here was something still more bizarre. For on 31 July, writing to Harlow Shapley,[30] he points out that it is just possible that the two supposed novae are in fact the same star which has altered position in the interval between the two photographs. Curtis, as a member of the Publication Committee of the ASP, tells Shapley he is "putting in a note on the nova in NGC 6946 [Ritchey's], and including also brief announcement of one I have found in NGC 4527, 14th mag., appeared in 1915; this makes four novae in spirals"; he goes on to describe his observations of NGC 4321 "which is either a case of *two* novae or a proper motion of 12" ± per year". Two days later he writes again to Shapley (in words which he incorporates in the *PASP* paper) to confirm that there are definitely two novae in NGC 4321.[31] "This now makes six novae which have appeared in spirals. Am inclined to believe that they must be novae and not variables from the evidence in the case of NGC 4527, where I have plates taken here or elsewhere covering the history of the region for seventeen years."

By October, when he discussed his results in a *Lick Observatory bulletin*,[32] Curtis had obtained the necessary plates of NGC 4321 and his account was as complete as he could have wished. The numbers of known novae were already increasing rapidly as other astronomers searched photographs of spirals,[33] and the anomalous brightness of the two nineteenth-century novae was to become

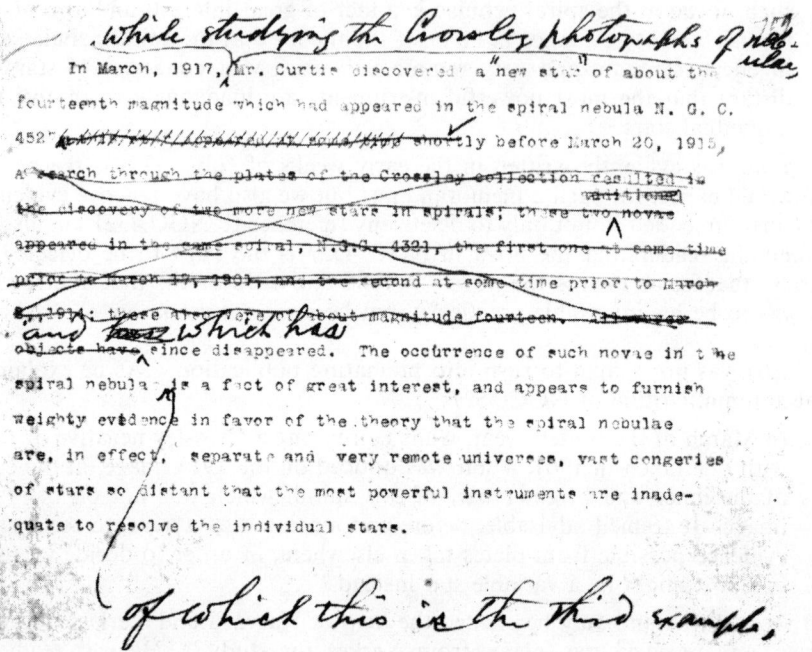

FIG. 1. Draft, amended in Curtis's hand, evidently for use in the Report on Lick Observatory for the year ended 30 June 1917 (courtesy Lick Observatory Archives).

steadily more apparent (indeed, in 1921 Curtis suggested[34] with insight: "a division into two magnitude classes is not impossible"). But even with only four new novae—his own three, plus the one subsequently discovered by Ritchey—he was already prepared to argue that the evidence now favoured the island-universe theory:

> The occurrence of these novae in spiral nebulae must be regarded as a fact bearing very directly on the theories of the constitution of the spirals. It seems to me that they furnish weighty evidence in favor of the well-known "island universe" theory of the spiral nebulae. The distribution of the novae thus far observed, excluding the six found in spirals, shows that they are essentially a galactic phenomenon; in fact, T *Coronae* is the only star amongst those called novae which lies outside the Milky Way, and T *Coronae* was not a typical nova. A limited number of novae—about twenty-six—have been found in our Galaxy. Though reasoning by analogy frequently leads to error, the occurrence of objects of the same type in the spirals would reasonably be expected, were these spirals in fact congeries of vast numbers of stars, like our own Galaxy.
> The entire invisibility of many novae before and after their brief apparition periods makes it possible to assume that they may well have increased in brightness at least sixteen magnitudes. On this rough assumption the novae in spirals would possibly have been of the thirtieth magnitude before their outburst. Stars of the fifteenth magnitude, in our own Galaxy,

if assumed to be twenty thousand light years away at present, would have a distance of the order of twenty million light years in order to be of the thirtieth magnitude. An external galaxy at this distance, if 10' in apparent diameter, would have an actual diameter of nearly sixty thousand light years, not an impossible dimension, so far as we may judge from our very imperfect knowledge of the size of our own Galaxy. Nebulae 10' in diameter are the giants of the class; the smaller spirals would have to be ten to one hundred times more remote, unless their dimensions were considerably smaller than our Galaxy is believed to be.[35]

The year 1917, then, is rightly remembered as that in which the study of novae and variables in spirals began in earnest; but the credit belongs first and foremost to Curtis. It was he who was engaged in a decade's campaign to study the spirals and detect changes in them; it was he who by mid-July had made a detailed study of one nova and the discovery of two more; it was he who proposed a challenging interpretation of the new evidence. The significance of Ritchey's discovery lay, not in the discovery as such, but in the fact that his nova was still bright and so could be comprehensively studied by astronomers the world over. Perhaps the resulting publicity did encourage more astronomers to search nebular photographs than would otherwise have been the case. But fundamental advances in the knowledge and understanding of novae in spirals were already assured well before the telegrams were despatched from Mount Wilson.

Acknowledgements

This paper results from a study of theories of the Galaxy and the nebulae *c.* 1920 carried out by the present writer and Professor Richard Berendzen. The kind co-operation of Mrs Grace Hubble, the archivist of the Shapley Archives at Harvard, and the directors of Lick and Lowell Observatories is gratefully acknowledged.

REFERENCES

1. See Richard Berendzen and Michael Hoskin, "Hubble's Announcement of Cepheids in Spiral Nebulae", *Astronomical Society of the Pacific leaflet* no. 504 (June 1971). Hubble's achievement was known to astronomers well before the formal announcement on 1 January 1925.
2. William Herschel, "Account of some Observations tending to investigate the Construction of the Heavens", *Philosophical transactions*, lxxiv (1784), 437–51. See M. A. Hoskin, *William Herschel and the construction of the heavens* (London, 1963), chap. 3.
3. See M. A. Hoskin, "Apparatus and Ideas in Mid-nineteenth-century Cosmology", *Vistas in astronomy*, ix (1967), 79–85, pp. 80–83.
4. N. U. Mayall, "Edwin Hubble, Observational Cosmologist", *Sky and telescope*, xiii (1953–54), 78–85.
5. Letter from Hubble to Harlow Shapley, 19 February 1924 (Hubble Archives, Huntington Library).
6. Kenneth Glyn Jones, "S. Andromedae, 1885", *JHA*, vii (1976), 27-40.
7. Agnes M. Clerke, *The system of the stars* (London, 1890), 370.
8. J. H. Reynolds writing on "Star Clusters and Nebulae" in *Hutchinson's Splendour of the heavens*, ed. by T. E. R. Phillips (2 vols, London, 1923), 571.
9. *Harvard College Observatory circular* no. 4 (20 December 1895).
10. *Ibid.*

11. V. M. Slipher, "The Radial Velocity of the Andromeda Nebula", *Lowell Observatory bulletin* no. 58 (1913). The result was communicated privately to a number of astronomers early in the year.
12. Letter of J. C. Duncan to V. M. Slipher, 17 February 1913 (Lowell Observatory Archives).
13. V. M. Slipher, "The Radial Velocity of the Andromeda Nebula", draft of *Lowell Observatory bulletin* no. 58 (Lowell Observatory Archives).
14. Edwin Hubble, *The realm of the nebulae* (New Haven, 1936), 85. Allan Sandage similarly gives priority to Ritchey in *The Hubble atlas of galaxies* (Washington, D.C., 1961), 2.
15. Hubble, *The realm of the nebulae*, 84.
16. Otto Struve and Velta Zebergs, *Astronomy of the twentieth century* (New York, 1962), p. 2 of endpapers.
17. *Harvard College Observatory bulletin* no. 641 (28 July 1917).
18. *Ibid.*, and H. D. Curtis, "New Stars in Spiral Nebulae", *Publications of the Astronomical Society of the Pacific*, xxix (1917), 180–2.
19. *Observatory*, xl (1917), facing p. 456.
20. Keeler's own programme was completed by C. D. Perrine, "Photographs of Nebulae and Clusters, Made with the Crossley Reflector, by James Edward Keeler, Director of the Lick Observatory, 1898–1900", *Publications of the Lick Observatory*, viii (1908), 1–46 with 70 plates.
21. Heber D. Curtis, "Changes in the Mounting of the Crossley Reflector", *Publications of the Astronomical Society of the Pacific*, xxii (1910), 40–41, and further papers in vols xxv (1913), 265–6 and xxvi (1914), 46–51.
22. H. D. Curtis, "Report of Work from July 1, 1912, to July 1, 1913", draft in Lick Observatory Archives.
23. H. D. Curtis, "Report 1913, July 1–1914, May 15", draft in Lick Observatory Archives. He refers to "the existence of exceedingly high radial velocities among the nebulae", and these may have decided Curtis in favour of the view that the spirals are galaxies.
24. "Edgewise or Greatly Elongated Spirals", draft list by Curtis, Lick Observatory Archives.
25. Lick Observatory Archives.
26. For example, his letter to Shapley of 31 July 1917 cited below.
27. Heber D. Curtis, "Three Novae in Spiral Nebulae", *Lick Observatory bulletin* no. 300 (drafted 8 September 1917 and issued 16 October), 108.
28. *Ibid.*, and Curtis, "New Stars in Spiral Nebulae" (ref. 18), 181.
29. Curtis, "New Stars in Spiral Nebulae" (ref. 18). The draft must have been amended to embody his further conclusions on NGC 4321, as the letters to Shapley discussed in this paragraph show.
30. Original in Shapley Archives, Harvard University.
31. Original in Shapley Archives, Harvard University. Curtis also wrote the same day to Slipher, informing him of his discovery of three novae and asking if Slipher had plates of the nebulae concerned (original in Lowell Observatory Archives); Slipher sent five plates of NGC 4321 but they were of little use as they were of only eight minutes' exposure. A third letter with the news of NGC 4321 was sent to E. C. Pickering at Harvard and incorporated in *Harvard College Observatory bulletin* no. 642 (9 August 1917).
32. Ref. 27. Correspondence between Campbell and Curtis now in the Lick Observatory Archives show that the Bulletin was written under difficulties as Curtis was away on war work at San Diego. The concluding paragraph, offering "another way" of establishing the distances of spirals from the novae contained in them, was inserted by Campbell with no more than acquiescence by Curtis.
33. See for example the reports in the October 1917 issue of *Publications of the Astronomical Society of the Pacific* (vol. xxix, no. 171), 210–17, and the remarks on novae by T. E. R. Phillips in "Report of the Council to the Ninety-eighth Annual General Meeting", *Monthly notices of the Royal Astronomical Society*, lxxviii (1917–18), 309. Phillips declares: "The Novae must almost certainly be *in* the nebulae, and this circumstance, taken in conjunction with the well-known clustering of Novae in our galaxy towards the galactic plane, seems highly significant, and gives striking support to the theory that the spirals are 'island universes' ".
34. In the published version of the so-called "Great Debate" (H. Shapley and H. D. Curtis, "The Scale of the Universe", *Bulletin of the National Research Council*, ii, Part 3 (May 1921), 215).
35. Curtis, "Three Novae in Spiral Nebulae" (ref. 27), 109.

5. *The 'Great Debate': What Really Happened*

The meeting of the National Academy of Sciences in Washington on 26 April 1920, at which Harlow Shapley of Mount Wilson and Heber D. Curtis of Lick Observatory both gave talks under the title "The Scale of the Universe", has passed into the literature as "The Great Debate".[1] It is true that the two resulting papers[2] published in the May 1921 *Bulletin of the National Research Council* contain the best presentations of the opposing arguments in the current controversy over the dimensions of our Galaxy and the status of the 'spiral' nebulae. But these papers, even if read without comment or discussion, would have taken well over two hours to deliver and therefore cannot possibly represent the proceedings at 'The Great Debate', which took place at 8·15 p.m. with a *Conversazione* timed to follow at 9·30.[3] Nevertheless, most historians persist in treating these published papers as the *verbatim* record of a dramatic trial of strength, and so have created an historical romance. In what follows we draw on surviving archives to compile a more accurate account of what actually took place.

The encounter grew out of a remark which George Ellery Hale, founder and Director of Mount Wilson Observatory, made at a Council Meeting[4] of the National Academy of Sciences late in 1919. Hale suggested that an evening of the Academy meeting planned for the following April should be devoted to one of the annual lectures paid from the fund set up in memory of Hale's father, William Ellery Hale.[5] On 3 January C. G. Abbot, the Home Secretary of the Academy, wrote to Hale:

> You mentioned the possibility of a sort of debate, either on the subject of the island universe or of relativity. From the way the English are rushing relativity in Nature and elsewhere it looks as if the subject would be done to death long before the meeting of the Academy, and perhaps your first proposal to try to get Campbell and Shapley to discuss the island universe would be more interesting. I have a sort of fear, however, that the people care so little about island universes, notwithstanding their vast extent, that unless the speakers took pains to make the subject very engaging the thing would fall flat. . . . Are there not other subjects—the cause of glacial periods, or some zoological or biological subject—which might make an interesting debate?[6]

It is a little surprising that the island universe theory of spiral nebulae—the claim that the spiral nebulae are galaxies in their own right and independent of our Milky Way star system—was to be defended, not by Curtis but by his Director at Lick Observatory, W. W. Campbell. For Curtis had been engaged for nearly a decade on the photography of nebulae with the Crossley reflector, and for much of that time had been an enthusiastic convert to the island universe theory; only that March he had dined with Hale in Washington within a week of lecturing on "Modern Theories of the Spiral Nebulae" to the Washington Academy of Sciences.[7] And on 8 October, when organizing the observing programmes for the 60in. and the new 100in. reflectors at Mount Wilson, Hale had written to Campbell to say "We are planning an extensive attack on spirals, with special reference to internal motion, proper motion, spectra of various regions, novae, etc., and here again *I should be glad to know what Curtis has in hand*, so that our work may fit in with it to advantage" (emphasis supplied).[8] Whatever the reason for the initial selection of Campbell as speaker,[9] by the time the question comes up again in correspondence Hale had received from Campbell a copy of the volume on nebulae published by Lick Observatory, in which "three splendid contributions"[10] were the work of Curtis, and thereafter Curtis and not Campbell is the projected speaker.

Meanwhile, however, Hale in fact favoured relativity, but on this Abbot had many misgivings:

> As to relativity, I must confess that I would rather have a subject in which there would be a half dozen members of the Academy competent enough to understand at least a few words of what the speakers were saying if we had a symposium upon it. I pray to God that the progress of science will send relativity to some region of space beyond the fourth dimension, from whence it may never return to plague us.[11]

Evidently Abbot's views prevailed, for he cabled Hale on 18 February: "Am wiring Heber Curtis suggesting Debate him and Shapley on subject scale of universe for Academy meeting forty five minutes each suggest communicate Shapley and Curtis and wire if favorably arranged."[12] Curtis accepted, at first with marked reluctance, then with increasing relish at the prospect of battle. Shapley likewise accepted—Hale was his 'boss' and the invitation a compliment —but with deep misgivings, for his career was now at a crossroads. In February 1919 the death of Edward C. Pickering had at last brought to a close his fortytwo-year reign as Director of the Harvard College Observatory. Pickering had been an outstanding administrator. The obvious choice as successor would have been Henry Norris Russell, Shapley's sometime teacher and mentor and the only American astronomer with influence comparable to that of Hale, except that he lacked Pickering's administrative abilities. Kapteyn, writing to Hale from Gröningen, thought Shapley perhaps the right candidate;[13] but Shapley, though a brilliant and original astronomer, was as yet only in his mid-thirties. This handicap did not deter Shapley. In later life he vividly recalled the day he heard of Pickering's death, and decided to "take a shot" at succeeding him.[14] He promptly wrote to both Russell and Hale to state his claims.[15] Russell was equally frank in reply: "To tell the naked truth, I would be very glad to see you in a good position at Harvard, free from executive cares. ... But I would *not* recommend you for Pickering's place, and I believe that you would make the mistake of your life if you tried to fill it."[16] To Hale, Russell remarked that Shapley "would not suffer if he pondered the old fairy tale about the man who got all sorts of good things from a magic fish whose life he had saved—until his wife wanted to be Pope!"[17] Hale warned Shapley: "My advice to any candidate for a position would be never to attempt to take an active part in securing it, as this is the surest way to defeat one's end."[18] Shapley, chastened but secretly unconvinced, wrote to both men to declare himself no longer a candidate.[19]

On 20 December, about the time that Abbot and Hale were considering Shapley for the Washington meeting, A. Lawrence Lowell, the President of Harvard, telegraphed to Mount Wilson: "Is Shapley coming East Xmas time for some scientific meeting? If so could he visit Cambridge? If not when could he come here?"[20] The secretary's reply that Shapley had no such plans led to a mysterious visit to Shapley from a Regent of Harvard. "He evidently sailed under sealed and secret and telegraphic orders," Shapley told Russell on 6 January with some excitement, "for he knew nothing of astronomy or physics or science, or me or anyone here. He asked about the scientific meetings here last June—that A.A.A.S. convention that I managed. . . . I might say that I am naturally very confident that Harvard is not too big for me and that the things I could and would do there would be a credit to American astronomy."[21] The visitor's interest in Shapley's ability as an organizer rather than as an astronomer was no doubt because he was being considered, not for the Directorship, but for a post in support of the new Director. Certainly in June Russell was to be offered the Directorship with "a second astronomer, younger, and with modern ideas, to be called, to act as the Director's right hand man" (Shapley was to Russell the obvious choice), and a *third* person to act as administrator;[22] and even when Russell eventually declined, Shapley was merely offered the post of Assistant Professor and Astronomer.[23]

Evidently believing he was nominated for the Directorship itself, and eager for the appointment, Shapley viewed the proposed encounter with Curtis with dismay. As ill-luck would have it, Curtis was an experienced and accomplished public speaker who might well put Shapley to rout, whatever the scientific merits of their respective cases, and this—taking place within easy reach of Harvard—could cost Shapley the Directorship. In the ensuing flurry of correspondence between Shapley, Curtis, Hale and Abbot, Shapley tried—half-heartedly—to get an Easterner substituted for Curtis,[24] and—tenaciously—to undermine the seriousness and length of the proposed encounter. Four distinguished and busy men repeatedly discussed whether it should be a 'debate' as originally proposed, or a 'discussion'—"two talks on the same subject from our different standpoints", as Shapley wished.[25] No sooner had Hale been won round by Shapley to a 'discussion' than the latter received from Curtis a letter which reawakened all his anxieties:

> I agree with you that it should not be made a formal "debate", but I am sure that we could be just as good friends if we did go at each other "hammer and tongs". . . . A good friendly "scrap" is an excellent thing once in a while; sort of clears up the atmosphere. It might be far more interesting both for us and our jury, to shake hands, metaphorically speaking, at the beginning and conclusion of our talks, but use our shillelahs in the interim to the best of our ability.[26]

Curtis sent a copy of the letter to Hale. It was 3 March before Hale could talk with Shapley and formulate his reply: "I do not think that the discussion should be called a 'debate', or that Shapley, who is perfectly willing to speak first, should have time allotted him for 'rebuttal'. If you or he wish to answer points made by the other, you can do so in the general discussion." Each should be manifestly a seeker after truth, "willing to point out the weak places in his argument and the need for more results".[27]

Not only had Shapley persuaded Hale away from the original concept of a debate, but he had convinced Hale that the proposed 45 minutes for each speaker was too long (on the grounds that this would tax the patience of the audience), and that 35 would be better. Curtis was aghast. The Lick Observatory Journal Club had recently devoted several meetings to the size of the Galaxy and the problem of external galaxies,[28] and Curtis had prepared a paper on the subject which he was circulating to friends for comment.[29] He knew how long he needed to make a serious scientific case. "We could scarcely get warmed up in 35 minutes", he protested to Hale.[30] Again the letters passed back and forth, and eventually a compromise of 40 minutes was imposed.[31]

The next problem concerned the subject matter. In Shapley's view, "The Scale of the Universe" made Curtis's main concern, the island universe theory, "an incident to a general discussion of the present guesses as to galactic dimensions and arrangement".[32] On the other hand, both men recognized that if, as Shapley maintained, the Galaxy was much larger than had previously been thought, it would be more difficult for Curtis to sustain the claim that the spiral nebulae were independent island universes; and it was clear from the pictorial slides Curtis proposed to use that he would indeed concentrate on the island universe theory.[33] Shapley welcomed their different but interrelated approaches as offering scope for a partnership instead of the dreaded confrontation: "I shall not be able to get as far as details of nebulae in my half of the talk, but I shall get some of the explanatory, introductory, illustratory requisites out of the way so that you can probably go farther into the details."[34] But he knew that Curtis planned to present a serious scientific argument, summarized on typewritten slides which he would show to the audience.[35] Shapley decided to appeal to Russell, his powerful ally, for vocal support, though putting the suggestion as usual into the mouth of a senior colleague:

> I lead off (with pictures), then Curtis presents his views, and then follows general discussion. Mr. Hale is anxious that you lead that discussion in whatever way you see fit, and I believe he plans to ask the presiding officer to call upon you as a starter. . . .
>
> Curtis swears by Newcomb and other patriarchs, and will show(?) that my distances are some ten times too big. Now that ten times, as Mr. Hale realizes, is as bad on your hypotheses as on mine; it is a violation of nearly all recent astrophysical theory. So unless Curtis actually bowls us over with the only true truth in these celestial matters, you will be interested in this general assault from the self-styled conservatives.
>
> Professor Brown is here at the observatory; also Professor Frost. They, as well as the people at Lick and at Mount Wilson, seem to regard that coming discussion as a crisis for the newer astrophysical theories. . . . But, crisis or not, I am requested to talk to the general public of non-scientists that may

happen to drop in. Consequently whatever answer must be made to Curtis and his school must be made in the discussion.

I write you this because you may be interested in knowing what the situation is, and so that you may be ready to defend your own views if they are imposed upon by either of us. To make matters worse for me, Mr. Agassiz of the Harvard Obs. Visiting Committee is coming down to the lecture and to eat a lunch with me; and A.L.L. himself has written for an appointment in Washington.[36]

In fact Russell made so substantial a contribution in support of Shapley that the question arose of whether he should be a third author of the published version, for in July Shapley told Curtis: "Russell is probably not coming in the published discussion, according to Hale, so either I should have the come-back or I should know what you are going to do and rebut in advance."[37]

There remained the crucial question of the content and level of Shapley's presentation in Washington. His decision was to treat the National Academy of Sciences to an address so elementary that much of it was necessarily uncontroversial. The typescript he used—covered with pencilled emendations, some in shorthand—runs to some 19 pages.[38] Of these, the last three pages are devoted to the intensifier he had developed to permit the photography of very faint stars —irrelevant to the theoretical argument, but perhaps directed in part to those members of the audience responsible for the future development of Harvard College Observatory. Of the first 16, it takes him more than six to reach the definition of a light year! The remaining ten pages are published below; this, and not the technical paper which appeared over a year later, was what Shapley actually said in Washington.

Although Curtis intended to present his case through a series of typewritten slides, he also had a script of sorts, no longer extant. It was probably by way of introduction to the more technical material on the slides, for he wrote to Shapley the following August: "I am sending with this a copy of my talk at Washington. This will recall to you the general lines of the arguments used. . . . Unfortunately, most of my actual argument was shown in the form of typewritten slides; I have no copy of these to send on to you at present. . . ."[39] These slides (or some of them) have survived and are reprinted below. They relate closely to the published account, and at Washington must have formed an odd contrast to the elementary talk by Shapley which had preceded. The contrast is echoed in Shapley's letter to Curtis on 9 June, telling him that Hale thought that "in a slightly different form the papers would go to the Proceedings—he favors that, in fact, even if the papers are long, providing the material is suitable in being not too popular (like mine?) or too tabular or technical (like yours?)". Curtis modestly accepted the criticism: "Yes, I guess mine was too technical. I thought yours would be along the same line, but you surprised me by making it far more general in character than I had expected. Had some thoughts of changing entire character of my presentation about five minutes before close of your part, but decided at last minute to go ahead with program as planned."[40]

A referee might have declared 'no contest', but insofar as there was a contest, Curtis was the winner. Shapley in old age recalled: "Now I would know how to dodge things a little better. . . . As I remember it, I read my paper and Curtis presented his paper, probably not reading much because he was an articulate person and was not scared."[41] Curtis, writing to his family on 15 May, reported "Debate went off fine in Washington, and I have been assured that I came out considerably in front".[42] Russell, writing to Hale in June about his invitation to become Director at Harvard, declared: "Shapley couldn't swing the thing alone. I am convinced of this after trying to measure myself with the job, and observing Shapley at Washington"—but if Shapley joined Russell there as his 'second' he ought to offer a lecture course for this "cultivates the gift of the gab, which he needs".[43]

In spite of the disparate performances, the occasion lived in the memory of those directly or peripherally involved.[44] For Curtis it was the climax of his decade of research on the nebulae; by July he was at Allegheny Observatory as Director and his creative years as an observer were over. For Shapley it was the occasion when he (and Mrs Shapley!) were vetted for the Harvard appointment. For the staff of the two great Californian observatories, it was something of a duel between champions.[45] Above all, the time was ripe for an appraisal of conflicting evidence and opposed interpretations on the fundamental question of the nature of the universe in the large; as R. G. Aitken of Lick remarked,

> I would like to hear the debate between Curtis and Shapley. I have read Curtis' paper—a very good one—and have had long talks with Shapley also, and each one has many very good arguments to present. For my own part, I am still "on the fence" on the question. I very greatly doubt the visibility of half-a-million or more 'island universes' on the one hand, and, on the other, I am not ready to accept Shapley's conclusions *on the basis of his measuring-rod*. It seems to me that its value is not yet sufficiently demonstrated. I am open to conviction.[46]

Curtis went prepared for the fur to fly;[47] his contribution was by common consent well presented and, as the slides show, at a high technical level. Russell, a talker of legendary capacity[48] and an outstanding astronomer, made a substantial reply from the floor, and we may be sure the rest of the session was hard fought. No wonder it was a memorable occasion. But the scientific argument and counterargument between Shapley and Curtis enshrined in the *Bulletin* papers belong, not to the verbal fisticuffs of Washington, but to their ensuing and protracted correspondence; and that is another story.

Acknowledgements

This paper results from a study of theories of the Galaxy and the nebulae c. 1920 carried out by the present writer and Professor Richard Berendzen. Dr N. E. Wagman, when Director of Allegheny Observatory, kindly supplied copies of the slides used by Curtis. Mrs Margaret C. Walters (née Curtis), Mrs Margaret Russell Edmonson, and the Shapley family generously allowed use of materials by Curtis, Russell and Shapley respectively. Grateful thanks are also due to Dr H. W. Babcock (Director, Hale Observatories), Dr Clark A. Elliott (Associate Curator, Harvard University Archives), Mr Alexander P. Clark (Curator of Manuscripts, Princeton University Library) and Miss Jean R. St. Clair (Archivist, National Academy of Sciences).

SHAPLEY'S WASHINGTON ADDRESS[49]

... Now that we have a satisfactory unit of sidereal distance [the light-year], let us go rambling about the universe. We see at one edge of this Milky Way field a cluster of stars—a typical phenomenon in the galactic system. To the unaided eye we see the Pleiades as such a cluster; stars of the constellation Orion also form a real physical system of stars moving together and probably all of common origin. We know from special investigations that even the big dipper is a stellar organization. We may, indeed, trace by continuous steps the denseness and richness of the clustering motive from the richest of globular clusters to the poorly organized nearby systems. Messier 11 is a society of a few hundred stars, forming an open cluster in a rich part of the galactic clouds. In the same region is Messier 22, a transition type from open to globular clusters. It is one of the nearest systems of its class—only 25,000 light-years away—nearer and brighter than the great cluster in Hercules, Messier 13, but not so well known because far south and less condensed. This eleven hour exposure of Messier 13, made by Ritchey with the 60-inch reflector, probably shows the faintest stars ever photographed with that telescope. Since all globular clusters are very much alike except in distance, this picture is suitable for an illustration of the dimensions and physical properties of a typical system.

We do not know how many stars are in Messier 13, probably not less than 50,000; about 30,000 have been counted, and the bottom has not yet been reached. The distance of the cluster, I find, is some 35,000 light-years; its linear diameter is therefore more than 350 light-years, and its total radiation is 300,000 times that of our sun.

The Hercules cluster has been extensively studied. We know for example the positions, magnitudes, and colors of all its brightest stars, and many relations between color, magnitude, distance from the center and star density. We now have the spectra of many of the individual stars, and their radial velocities; and the velocity and spectral type of the cluster as a whole. We know the types and periods of light variation of its variable stars, the colors and spectral types of these variables, and something also of the absolute luminosity of the brightest stars of the cluster from the appearance of their spectral lines. With knowledge of all this structural and historical detail, is it surprising that we venture to determine the distance of Messier 13 and similar systems with more confidence than was possible ten years ago when not a one of these facts was known or seriously thought about?

I shall not impose upon you the dreary technicalities of the methods of determining the distance of globular clusters. That would involve discussion of parallactic motion, probable errors, Cepheid variation, giant stars, and such matters. I think it will suffice to sketch briefly the principles involved.

For nearby stellar objects we can make direct trigonometric measures of distance, using the earth's orbit or the path of the sun through space as a base line. For many of the more distant stars the spectroscopic method is available, using the appearance of the stellar spectra and the readily measurable brightness of the stars; for certain types of stars, too distant for spectroscopic data, there is still a chance of obtaining the distance by means of the photometric method. This simple device, which is one particularly suited to studies of globular clusters, consists in determining, by some indirect means, the real light power of a star, that is, its so-called *absolute* magnitude, and then measuring its apparent magnitude. Obviously, if a star of known real brightness is moved away to greater and greater distances, its apparent brightness decreases; hence, for such stars of known absolute magnitude, the apparent magnitude gives, by a simple formula, the distance from the observer.

As I have suggested before, it is because within recent years we have advanced our knowledge so greatly that we can use these powerful spectroscopic and photometric methods of measuring distance. The advance is in two directions; first, in the study of the nearby stars we have learned of the uniformity in absolute brightness and spectroscopic characteristics of various types; and second, we have shown that in the clusters we have exactly the same kinds of stars as those around the sun—the same except that the cluster stars *appear* to be very faint. This apparent faintness of the cluster stars is due to *distance*, and is a *measure* of it. For instance, the very extensive studies by Plummer, Kapteyn, and Charlier, have shown that stars of spectral type B in the solar neighbourhood (the blue stars) are closely restricted around an average absolute luminosity about 200 times that of the sun. In Messier 13 we have the same types of stars. This we know from direct observations of spectrum with the 60-inch and 100-inch telescopes, supplemented by thousands of measures of color. But these blue cluster stars, which are actually about 200 times as bright as the sun, appear to be only one five-thousandth as bright as the sun would be if it were 33 light-years away; the distance of the blue cluster stars is therefore readily found to be some 35,000 light-years, and *their* distance is of course that of the whole cluster.

You may ask, however, is it not likely that these stars in the cluster, just because they are in a globular cluster, are of a different magnitude from our standards, even though comparable in color and spectrum? May they not be dwarfs in luminosity, and therefore not so far away? All the evidence, theoretical and observational, argues to the contrary. There should be little doubt in this matter of comparability for several good reasons. For instance, one reason—an all important one—*is that these nearby standard stars are themselves cluster stars*. I have already pointed out the continuous sequence from the densest globular cluster to the constellation Orion; and the stars in such open nearby clusters as the Hyades, Orion, Scorpio, *are* the standards.

It is thus because of this comparability of local B stars with those of the distant open and globular clusters, and because the absolute luminosities of the local stars are based upon thousands of good measures of proper motion and radial velocity, that the blue stars give us the strongest evidence for the great distances of globular clusters.

Similarly the giant red stars of our local clusters are found to be comparable with the many red stars in globular clusters, and qualitatively at least, they give through the photometric method the same distances for remote systems as the blue stars give.

Another class of stars, the Cepheid variables, have been used extensively, and in much the same way, in exploring not only the system of globular clusters but the star clouds of the Milky Way. By determining the light-curve of such a star in any cluster, the distance may be known with remarkable accuracy. The particular luminosity that corresponds to a given period is found, as for the B stars and red giants, from studies of nearby examples of the class. Professor Curtis may tell you more of the photometric method of getting at the distances of the nearby Cepheid variables—he may question the sufficiency of the data or the accuracy of the methods of using it. But this fact remains: we could discard the Cepheids altogether, use instead the thousands of B-type stars upon which the most capable stellar astronomers have worked for years, and derive just the same distance for the Hercules cluster, and for the other clusters, and obtain consequently the same dimensions for the galactic system.

To conclude, in the face of these results, that the Hercules cluster is not approximately at the distance derived for it photometrically, is to avoid the most direct, and simple, and conservative interpretation of the data. To suggest, as I believe Prof. Curtis may do, that the clusters are only one-tenth as remote as I place them, is equivalent to subscribing to views so radical in several departments of astronomy and physics that we instinctively hesitate.

If the distances I have assigned must be decreased to one-tenth, then the light-emitting power of distant cluster stars must be only *one-hundredth* that of local cluster stars *of exactly the same types*. As a consequence I believe Russell's illuminative theory of spectral evolution must be largely abandoned, and Eddington's brilliant theory of gaseous giant stars must be greatly modified or given up entirely. Now both of these modern theories have their justification, first in the fundamental nature of their physical concepts, and second, in their great success in fitting observational fact. Similarly, the period-luminosity law of Cepheid variation would be meaningless; Kapteyn's classic researches on the structure of the local cluster would need new interpretation, because his luminosity laws could be applied locally but not generally; and a very serious loss to astronomy would be that of the generality of the spectroscopic method of determining star distances, in fact, the whole application of that method as an independent tool, for it would mean that identical spectral characteristics may indicate stars differing by five magnitudes, depending merely upon whether the star is in the solar neighborhood or in a distant cluster.

With so many satisfactory methods and theories at stake—the very foundations of modern astrophysics—is it any wonder that we hesitate and argue against revolutionary interpretations? I believe I cannot follow Professor Curtis in calling those older, long-established interpretations conservative.

Suppose, therefore, we admit the obvious comparability of local cluster stars with those of distant clusters; is there not, however, a possibility that the distant stars appear faint through the loss of light in travelling through space? The Mount Wilson photometric studies show, however, no effect on star colors of such loss—a result checked by Hertzsprung and others. Two Swedish students find no suggestion of it in their studies of the very distant spiral nebulae. In the plane of the Milky Way, to be sure, we have dark nebulosity that *may* obscure more distant objects without affecting the color. But outside the galactic regions, and Messier 13 and most of the other globular clusters are far outside, we have in the distinct correlation of the angular size of a cluster with the brightness of its stars a fairly good proof of the absence of general light absorption. That is, if a cluster were removed to twice the present distance, its stars would be fainter, and also its area would be one-fourth as large. That is just what we observe—the faint clusters small, the small clusters faint—an obvious indication that the small faint clusters owe both of these characters to distance. Whereas, if the faintness of the cluster were due to obstruction of light, we should expect the angular diameter to be little affected; moreover, we should also expect to find, in such obstructed clusters, wholesale irregular variations, lop-sidedness, and other curious effects that are not observed, unless, miraculously, the hypothetical obstructing matter were exactly at rest with respect to the cluster beyond, or exactly uniform.

When we accept that the distance of the Hercules cluster is such that its stellar phenomena are all harmonious with local stellar phenomena, then it follows that fainter, smaller clusters are still more distant. Thirty of the 86 known are more distant than 100,000 light-years; the most distant is more than 200,000 light-years away, and the diameter of the whole system of globular clusters is about 300,000 light-years. Since the affiliation of the globular clusters with the Galaxy is shown by their concentration to the plane of the Milky Way and their symmetrical arrangement with respect to it, it also follows that the galactic system of stars is as large as this subordinate part. During the past year we have found Cepheid variables and other stars of high luminosity among the fifteenth magnitude stars of the galactic clouds; this can only mean that some parts of the clouds are more distant than the Hercules cluster. There seems to be good reason, therefore, to believe that the star-populated regions of the galactic system extend at least as far as the globular clusters.

One consequence of the cluster theory of the galactic system is that the sun is found to be very distant from the center of the Galaxy. It appears that we are near the center of a large local cluster or cloud of stars, but that cloud is at least 60,000 light years from the galactic center. Twenty years ago Newcomb remarked that the sun *appeared* to be in the galactic plane because the Milky Way is a great circle—an encircling band of light—and that the sun also *appears* near the center of the universe because the star density falls off with distance in all directions. But he concluded as follows:

"Ptolemy showed by evidence, which, from his standpoint, looked as sound as that which we have cited, that the earth was fixed in the center of the universe. May we not be the victim of some fallacy, as he was?"

The answer to Newcomb's question is: Yes, we have been victimized by the chance position of the sun near the center of a subordinate system, and misled by the consequent phenomena, to think that we are God's own appointed, right in the thick of things. In much the same way ancient man was misled, by the rotation of the earth, and by the consequent apparent daily motion of all heavenly bodies around the earth, to believe that even his little planet was the center of the universe, and that his earthly gods created and judged the whole.

If man had reached his present intellectual position in a later geological era, he might not have been led to these vain conceits concerning his position in the physical universe, for the solar system is going rapidly away from the center of the local cluster. If that motion remains unaltered in direction and amount, in a hundred million years or so the Milky Way will be quite different from an encircling band of star clouds, the local cluster will be a distant object, and the star density will no longer decrease with distance from the sun in all directions.

Remembering these delusions, relative to his physical status in the universe, may we not appropriately ask if man is also biologically blindfolded? Does he, perhaps, hold his self-assumed and self-defined position at the peak of animal development as a victim of psychological fallacy?

Another consequence of the conclusion that the galactic system is 300,000 light-years or more in greatest diameter, is its bearing on the problem of the spiral nebulae. I shall leave the description and discussion of this debatable question to Professor Curtis. We agree, I believe, that if the galactic system is as large as I maintain, the spiral nebulae can hardly be comparable galactic systems; if it is but one-tenth as large, there *might* be a good opportunity for the hypothesis that our galactic system is a spiral nebula, comparable in size with the other spiral nebulae, all of which would then be "island" universes of stars. On one other point I think we also agree, or at least we should agree, and that is that we know relatively so little concerning the spiral nebulae and we are soon going to know relatively much because of the increasing activity in the nebular field, that it is professionally and scientifically unwise to take any very positive view in the matter just now.

But to summarize my view, which I hope is not positive and which is certainly subject to change if future data justify, the spiral nebulae are distant objects, not members of our galactic system, except that the nearer brighter ones have some sort of a relation to the Galaxy, not only in distribution, but also in motion. This relation to the Galaxy may be transitory for any given nebula, for with their enormous speeds of recession, they may eventually pass out of our domain of space. I prefer to believe that they are not composed of stars at all, but are truly nebulous objects. For instance, these two photographs of the typical spiral, Messier 51, by Mr. Seares indicates that the nebulosity is not composed of stars—as required by the island universe theory. The picture on the left is taken with a plate sensitive to yellow light, the one on the right records the blue light. The exposures are adjusted so that the superposed stars are of the same intensity. If the arms of the spiral were composed of stars the two pictures should be comparable in intensity throughout. If, however, the spiral is truly nebulous, it would appear faint on the photograph in yellow light. That, you see, is the actual condition in this system. No type of star is known with anything like as large a negative color index as is shown by the nebulosity in this spiral. But even if the spirals are stellar, they are not comparable in size with our stellar system, and our system is not comparable in constitution with the spiral nebulae. Professor Curtis, I hope, will have time to go farther into this interesting question. . . .

SLIDES SHOWN BY CURTIS IN WASHINGTON[50]

[A] THE SIZE OF OUR GALAXY; OLDER VIEW

Studies of the distribution of the stars and the ratios between the numbers of stars of successive magnitude have led a number of investigators to fairly accordant dimensions for the galaxy.

Wolf; about 14,000 light-years in diameter.
Eddington; about 15,000 light-years.
Shapley (1915); 20,000 light-years.
> "That the maximum radius of the Milky Way is probably not greater than ten thousand light-years and may be somewhat less has been deduced from many lines of evidence, the most important of which is the color of the faint stars." (Mt. Wilson Contr., No. 116, 1915.)

Newcomb; not less than 7,000 light-years; later—perhaps 30,000 light-years in diameter and 5,000 light-years in thickness.

A maximum galactic diameter of 30,000 light-years will be assumed as representing sufficiently well the older view; it is perhaps too large.

[cf. ref. 2, 195–6]

[B] THE SHAPE OF OUR GALAXY OF STARS

Studies of the distribution of the stars over the entire sky, with investigations based on the ratios between the numbers of stars of successive magnitudes, have given the following results:

1. The stars are not infinite in number, nor uniform in distribution.
2. Our Galaxy, delimited for us by the projected contours of the Milky Way, contains possibly a billion suns.
3. Our Galaxy is shaped much like a lens, or a thin watch, the thickness being perhaps one-sixth of the diameter.
4. Our sun is located fairly close to the center of figure of the Galaxy.
5. The stars are not distributed uniformly through this galaxy. A large proportion may be actually in the ring structure suggested by the appearance of the Milky Way. There is some slight evidence for a spiral structure. Our position near the center of figure of the Galaxy is not a favorable one for a determination of the actual galactic structure.

[cf. ref. 2, 196]

[C] THE SIZE OF OUR GALAXY; SHAPLEY'S VIEW

From evidence to be referred to more fully later, Dr. Shapley has derived very great distances for the globular star clusters, 220,000 light-years for the most remote.

The apparent distribution of these globular clusters shows incontrovertibly that they are an integral feature of our galactic system.

This evidence has formed the main reason for Dr. Shapley's adoption of a diameter of 300,000 light-years for our galactic system, fully ten times greater than that accepted hitherto.

[cf. ref. 2, 197–8]

[D] THE STARS OF THE MILKY WAY

The smaller postulated dimensions for the Galaxy require stars whose absolute magnitudes are in fair accord with those of known distance. The larger dimensions require a very large proportion of giant stars.

Apparent Magnitudes	Corresponding absolute magnitudes for distances of	
	10,000 l.y.	100,000 l.y.
8	−4·4	−9·4
10	−2·4	−7·4
12	−0·4	−5·4
14	+1·6	−3·4
16	+3·6	−1·4
18	+5·6	+0·6
20	+7·6	+2·6

[cf. ref. 2, 200]

[E]

The conditions of star concentration obtaining in the Magellanic Clouds and in the globular clusters appear to render these regions of space unique as regards variable stars.

The Magellanic Clouds contain .. .. 1800 variable stars

Total of all variables in the rest of the sky, excluding those in globular clusters 1686

The globular clusters contain numbers of variable stars ranging from 137 in N.G.C. 5272 to 0 for N.G.C. 3293 and 4755. Practically all are shorter than one day in period. Total .. 509

Short period cluster-type variables discovered to date in the rest of the sky 45

[?Replaced in ref. 2 by discussion of Cepheid variables and "giant" stars]

[F] THE SPECTRUM OF THE SPIRAL NEBULAE

As island universes	As galactic phenomena
The spectrum of the average spiral nebula is indistinguishable from that given by a star cluster.	If the spiral nebulae are an integral part of our Galaxy, we must assume that they are some sort of finely divided matter, or of gaseous constitution.
It is such a spectrum as would be expected from a vast congeries of stars.	
In general type it resembles the integrated spectrum of our Milky Way.	If galactic, we have no adequate and actually existing evidence by which we may explain their spectrum.
The spectrum of the spiral nebulae offers no difficulties in the island universe theory of the spirals.	The diffuse nebulosities of our galaxy give a bright-line gaseous spectrum. A few, associated with bright stars, agree with their involved stars in spectrum, and are well explained as a reflection or resonance effect.
	Such an explanation is untenable in the case of a large proportion of the spirals.

[cf. ref. 2, 212]

[G] THE DISTRIBUTION OF THE SPIRAL NEBULAE

The spiral nebulae are found in greatest numbers just where the stars are fewest (at the poles of our Galaxy), and not at all where the stars are most numerous (in our galactic plane). No spiral has as yet been found actually within the structure of the Milky Way.

As island universes

It is most improbable that our galaxy should, by mere chance, be placed about half-way between two great groups of island universes.

So many of the edgewise spirals show peripheral rings of occulting matter that this dark ring may be the rule, rather than the exception.

If our Galaxy, itself a spiral on the island universe theory, possesses such a peripheral ring of occulting matter, this would obliterate the distant spirals in our galactic plane, and explain their peculiar distribution.

There is some evidence of such occulting matter in our galaxy.

Additional observations on the spirals south of the galactic plane may remove this recession excess. Part of this may also be due to the motion of our Galaxy in space.

As galactic phenomena

If the spirals are galactic objects, they must be a class apart from all other known types.

Their abhorrence of the regions of greatest star density can only be explained on the hypothesis that they are, in some manner, repelled by our Galaxy.

We know of no force adequate to produce such a repulsion, except perhaps light pressure.

Why should this repulsion invariably have acted at right angles to our galactic plane?

Why have not some been repelled in the direction of our galactic plane?

The repulsion theory is given some support by the fact that most of the spirals observed to date are receding from us.

[*cf.* ref. 2, 213]

[H]   NEW STARS IN THE SPIRAL NEBULAE (1)

Within the past few years some twenty-five novae have been discovered in spiral nebulae, sixteen of these in the Nebula of Andromeda, as against about thirty in historical times within our own galaxy.

Apparent Magnitudes

	Thirty galactic Novae	Seventeen Novae in Neb. Andromeda
At maximum	+ 5	about +17
At minimum	+15	perhaps +27 ?

Absolute Magnitudes

	Novae in Nebula of Andromeda, if at distance of		Four galactic Novae of known distance
	20,000 l.y.	500,000 l.y.	
At maximum	+ 3·1	−3·9	−3·4
At minimum	+13·1	+6·1	+7·2

[*cf.* ref. 2, 214–15]

[J]   THE SPIRAL NEBULAE AS ISLAND UNIVERSES
Summary

1. On this theory we avoid the almost insuperable difficulties involved in the attempt to place the spirals in any coherent scheme of stellar evolution, either as a point of origin, or as a final evolutionary product.

2. On this theory, it is unnecessary to attempt to coordinate the tremendous space-velocities of the spirals with average star velocities.

3. The spectrum of the spirals is like that given by a star cluster.

4. A spiral structure for our own Galaxy has been suggested, and is not improbable.

5. If island universes, the new stars observed in the spirals seem a natural consequence of their nature as galaxies. Correlations between the new stars in spirals and those in our Galaxy indicate a distance ranging from perhaps 500,000 light-years in the case of the Nebula of Andromeda, to 10,000,000, or more light-years for the more remote spirals.

6. At such distances, these island universes would be of the order of size of our own Galaxy of stars.

7. Very many spirals show evidence of peripheral rings of occulting matter in their equatorial planes. Such a phenomenon in our own Galaxy, regarded as a spiral, would serve to obliterate the spirals near our galactic plane, and would furnish an adequate explanation of the peculiar distribution of the spiral nebulae.

[*cf.* ref. 2, 216–17]

REFERENCES

1. Discussions include: Otto Struve, "A Historic Debate about the Universe", *Sky and telescope*, xix (1959–60), 398–401; Norriss S. Hetherington, "The Shapley–Curtis Debate", Astronomical Society of the Pacific Leaflet no. 490 (April 1970); Otto Struve and Velta Zebergs, *Astronomy of the twentieth century* (New York, 1962), chaps 19 and 20, *passim*; and "The Great Debate", chap. 6 of Harlow Shapley, *Through rugged ways to the stars* (New York, 1969).
2. H. Shapley and H. D. Curtis, "The Scale of the Universe", *Bulletin of the National Research Council*, ii, Part 3 (May 1921).
3. As shown by the official programme of the Academy meeting.
4. The meeting took place on 19 December. There is no reference in the minutes to Hale's suggestion.
5. W. E. Hale had used his wealth to support his son's projects, notably by providing the disc for the 60in. telescope eventually erected at Mount Wilson.
6. Archives of the National Academy of Sciences.
7. As shown by the letters from Curtis to his children, 8 February and 9 March 1919 (Michigan Historical Collections, University of Michigan). The script of Curtis's lecture is in the archives of Lick Observatory.
8. Hale microfilm.
9. Not surprisingly, in 1968 Dr Abbot did not recall the reason for the choice of Campbell, but remarked that Campbell was of course "a more important astronomer than Curtis" (personal communication). Mr Robert Smith points out that Campbell had supported the island universe theory in "The Nebulae", *Science*, xlv (1917), 513–48.
10. Letter of Hale to Curtis, 24 February 1920 (Hale microfilm).
11. Letter of Abbot to Hale, 20 January 1920 (Hale microfilm).
12. Hale microfilm. Hale cabled at once to Shapley and Curtis, offering each an honorarium of $150.
13. Letter of 7 February 1919 (Hale microfilm):
    In America you have Russell and Shapley. Shapley is a brilliant man and personally I, who know him mainly only through his scientific work, would think him the best fitted for the position. Meanwhile I do not know him sufficiently to know how he would do as an organiser at the head of such a large and complicated Institution as the Harvard Observatory.
14. Shapley, *Through rugged ways to the stars*, 82.
15. Shapley to Russell, 12 February 1919; to Hale, 13 February 1919 (Shapley Archives, Harvard University).
16. Russell to Shapley, 19 February 1919 (Shapley Archives, Harvard University).
17. Russell to Hale, 19 February 1919 (Russell Archives, Princeton University).
18. Hale to Shapley, 27 February 1919 (Hale microfilm).
19. Shapley to Russell, 27 February 1919; to Hale, 7 March 1919 (Shapley Archives, Harvard University).
20. Hale microfilm.
21. Shapley to Russell, 6 January 1920 (Russell Archives, Princeton University).
22. Russell to Hale, 13 June 1920 (Hale microfilm).
23. Invitation dated 10 November 1920 (Shapley Archives, Harvard University).
24. Shapley to Hale, "Sunday", *i.e.* 22 February 1920 (Hale microfilm).
25. Shapley to Hale, 19 February 1920 (Hale microfilm). On the 24th Hale wrote to Shapley, Curtis and Abbot approving the concept of a 'discussion'.
26. Curtis to Shapley, 26 February 1920 (Shapley Archives, Harvard University).
27. Hale to Curtis, 3 March 1920 (Shapley Archives, Harvard University).

28. Curtis to E. E. Barnard, 28 January 1920 (Archives of Yerkes Observatory). Curtis reports that "most of us here find it impossible to subscribe to some of the recent theories on these points".
29. *Ibid.* On 23 February Curtis requested from Barnard the return of this paper as a matter of urgency.
30. Curtis to Hale, 9 March 1920 (Hale microfilm).
31. Abbot to Hale, 18 March 1920 (Hale microfilm).
32. Shapley to Abbot, 12 March 1920 (Shapley Archives, Harvard University).
33. Curtis to Shapley, 14 March 1920. Since Shapley maintained to the end of his life that Curtis did not address himself to the subject of the title (Shapley, *Through rugged ways to the stars*, 79), it is worth recording the synopsis Curtis proposed to Hale (Curtis to Hale, 20 February 1920, Shapley Archives, Harvard University) and which Hale welcomed: "Dr Shapley will discuss recently secured evidence pointing to dimensions of our galaxy about ten times greater than held in the older theories of the Milky Way, i.e., a diameter of about 300,000 light-years, with the spiral nebulae regarded as a galactic phenomenon. Dr Curtis will defend the older view that our Milky Way is approximately of the dimensions suggested by Newcomb, i.e., about 30,000 light-years in diameter, with the spiral nebulae regarded as very probably individual galaxies, or 'island universes'."
34. Shapley to Curtis, 18 March 1920 (Shapley Archives, Harvard University).
35. Curtis to Shapley, 14 March 1920 (Shapley Archives, Harvard University).
36. Shapley to Russell, 31 March 1920 (Shapley Archives, Harvard University).
37. Shapley to Curtis, 27 July 1920 (Shapley Archives, Harvard University).
38. Shapley Archives, Harvard University.
39. Curtis to Shapley, 2 August 1920 (Shapley Archives, Harvard University).
40. Shapley to Curtis, 9 June 1920; Curtis to Shapley, 13 June 1920 (Shapley Archives, Harvard University).
41. Shapley, *Through rugged ways to the stars*, 79–80. In private conversation Shapley was much more emphatic as to his disappointing performance.
42. Michigan Historical Collections, University of Michigan.
43. Russell to Hale, 13 June 1920 (Hale microfilm).
44. Writing to Shapley on 10 July 1922, Curtis spoke of "our memorable set-to" (Archives of Allegheny Observatory). C. D. Shane of the University of California (Berkeley) wrote to Curtis on 3 December 1923 about "the famous debate", and Curtis in reply on the 10th again referred to "our memorable set-to" (Michigan University Archives). Campbell, writing on "Do we live in a spiral nebula?" in *Popular astronomy* for 1926, speaks of the "memorable discussion" of 1920 (p. 175).
45. Robert G. Aitken later wrote of Curtis that "For a time only his colleagues at Mount Hamilton, and a few other astronomers agreed with him in his views" (*National Academy of Sciences Biographical Memoirs*, xxii (1943), 280), and this may be true, although Aitken had forgotten that he himself had been "on the fence" (*cf.* ref. 46 below).
46. Aitken to Barnard, 27 April 1920 (Archives of Yerkes Observatory); emphasis in original.
47. "... some fur ought to fly, on both sides", Curtis to W. J. Hussey, formerly of Lick Observatory, 15 April 1920 (Michigan Historical Collections, University of Michigan).
48. "He sure is a talker.... I never saw any man better qualified to teach the unwashed astronomy then he", "Benny" writing to Shapley of a talk by Russell to the general public at Mount Wilson, 30 June 1921 (Shapley Archives, Harvard University).
49. This rough typescript, now in Box 1 of the Shapley Archives at Harvard University, contains pencil amendments in longhand, and occasionally in shorthand. Those in longhand have been incorporated in this printed text. Of the typescript, the first one-third is too elementary to justify reprinting; likewise, the final three pages dealt with Shapley's intensifier, and, however significant this might have been as an instrumental advance in the study of faint stars, it was not directly relevant to the theoretical discussion and is omitted here as it was in the printed version of the proceedings (ref. 2).
50. The slides survive at Allegheny Observatory. Of the nine, eight are represented in modified form in the printed version of the proceedings, as indicated, and the slides have accordingly been arranged in a probable order. The title of slide H suggests that Curtis may have used other slides no longer extant, but surely very few additional slides could have been fitted into a 40-minute talk.

Acknowledgements

Nine of the chapters of this book are detailed studies that first appeared in the pages of *Journal for the history of astronomy* and are here reprinted with minor changes and additions. These chapters are: Section A, chaps. 4 and 5; Section B, chaps. 2, 3 and 4 (but the Postscript to chap. 2, on Halley and 'Olbers's Paradox', is new); and Section C, chaps. 1, 4 and 5. The remaining two detailed studies are in Section A: chap. 2 appeared in *Sudhoffs Archiv,* published by Franz Steiner Verlag of Wiesbaden, West Germany, and chap. 3 is published for the first time.

The remaining chapters are intended to set these detailed studies in context. The Introduction, on "Principles and Methods", is a second attempt at this theme, the first having appeared in *Human implications of scientific advance,* ed. by E. G. Forbes (Edinburgh, 1978). Section A, chap. 1 is based on an article in *Avant, avec, après Copernic* (Paris, 1975), published by Albert Blanchard. Section B, chap. 1, and Section C, chaps. 2 and 3, are new, but I have written on these subjects before and have here cannibalised an occasional paragraph or two.

Index

Abbot, C. G.: 175, 176, 177
Aberration of light: 3, 7-8, 12, 33-35
Agassiz, G. R.: 179
Airy, G. B.: 143
Aitken, R. G.: 180
Aldebaran (α Tau): 8, 59, 60, 65, 66
Algol (β Per): 13, 25, 26, 37, 40-44, 46, 49, 51
Altair (α Aqu): 8, 56, 61
Andromeda Nebula (M31)
   Cepheid variables in: 150-1, 160-4, 168
   distance of: 161, 164, 168
   novae in: 156, 157, 168, 169 (S Andromedae); 160-1, 186
   observed by W. Herschel: 126, 128-9, 131
   radial velocity of: 157, 169
   seventeenth-century observations of: 24, 25, 125
   spectrum of: 1, 155
   status of: 139, 143, 150, 154, 155
S Andromedae: see Andromeda Nebula
Annual parallax, measurement of: 2, 5-6, 8-11, 29, 75, 80, 88, 117, 181
   double-star method of: 9-11, 12, 14, 92 n24
   by observation of 61 Cygni: 9-11
   by observation of Pole Star: 7, 8
   by observation of star in zenith: 7, 8, 29-34, 35
   see stars, nearest, criteria for
Anthelme, V.: 25, 26
η Antinoi: see η Aquilae
α Aquilae: see Altair
η Aquilae (Antinoi): 45, 46-47, 49, 51
Arago, D. F. J.: 9
Arcturus (α Boo): 8, 38, 56, 61, 63, 65
Argelander, F. W. A.: 12, 14, 16, 52
γ Arietis: 59, 60, 61
Astronomical unit: 5, 33, 34-35, 76, 114, 115
Atmospheric refraction, effect of Earth's motion on: 32-33
Aubert, A.: 63
α Aurigae: 8
Auwers, A. J. G. F. von: 12

Babcock, H. W.: 180
Banks, J.: 43, 44, 50, 51
Barchas, S. I.: 114

Bayer, J.: 24, 40
Bentley, R.: 67, 72-74, 78, 87, 88, 119
Berendzen, R. B.: 173, 180
Berge, M.: 8
Bessel, F. W.
   and annual parallax: 5, 9-11
   and dark companions of Procyon and Sirius: 15
   Fundamenta astronomiae: 12, 14, 16, 34
   and solar motion: 14
Bird, J.: 37
Black holes: 154, 166
Blaeu, W. J. (Jansonius): 23, 24
α Bootis: see Arcturus
Boulliau, I.: see Bullialdus
Boyle, R.: 72
Bradley, J.:
   and discovery of aberration: 3, 7-8, 12, 31-35
   and double-star method of measuring annual parallax: 9
   his observing programme: 12
   and proper motions: 13, 119
   and stellar distances: 5, 34-35
Brahe: see Tycho Brahe
Brieux, A.: 117
Brinkley: J.: 8-9
Brown, E. W.: 178
Bullialdus, I. (Boulliau): 13, 24, 25, 37
Bürgi, J.: 24

Calandrelli, G.: 8, 9
Campbell, T.: 3, 154
Campbell, W. W.: 175, 176
α Canis Majoris: see Sirius
α Canis Minoris: see Procyon
Capella (α Aur): 8
Carochez, N.-S.: 136
Cassini, G. D. (Cassini I): 25
Cassini, J.-D. (Cassini IV): 136, 137
α Cassiopeiae: 44, 51
Castor (α Gem): 61
Catalogues
   of comparative brightnesses of stars: 12, 19 n52, 44, 49-50
   of nebulae: 125, 127 (Halley); 125-6, 127,

128, 129, 134 (Messier); 134, 137 (W. Herschel); 143, 145 (J. Herschel)
  of stars: 11-12, 16, 34; *see also under* Flamsteed
  of variable stars: 37, 48
α Centauri: 11
Z Centauri: 168-9
δ Cephei: 45, 46, 47, 49, 51
Cepheid variables: 6, 47, 150-1, 160-4, 169, 180, 182, 183, 185; *see* η Aquilae; δ Cephei
o Ceti: *see* Mira
Charlier, C. V. L.: 181
Chéseaux, J. P. L. de: 96
Circles, by Ramsden: 8
Clark, A. G.: 15
Clark, A. P.: 180
Clarke, S.: 88
Comets: 37, 38, 39, 44, 51, 86, 118
  collisions with stars as cause of novae: 13
  habitable (Whiston): 112
  Halley's: 11
  W. Herschel's theory of: 140-1
  location of Hell: 102
  Wright's theory of: 112
Copernican hypothesis proved by aberration: 8, 31
Copernicus, N.: 5, 29
R Coronae Borealis: 48-49, 51
T Coronae Borealis: 172
Cosmogony
  of W. Herschel: 122, 131, 133, 134, 135, 139, 141, 142
  of Kant: 68-69
Cotes, R.: 87
Crab Nebula (M1): 149, 150
Curtis, H. D.: 157, 169-73, 175-87
α Cygni: 8
β Cygni: 59, 60, 61, 62, 63
ε Cygni, 61, 62, 63
χ Cygni: 37, 38, 39, 40, 41, 44, 49
61 Cygni: 2, 9-11, 18

Darquier de Pellepoix, A.: 117
D'Arrest, H. L.: 151
Delambre, J.-B. J.: 50
Deneb (α Cyg): 8
Derham, W.: 67-68
Distances: 181
  of Sun: *see* astronomical unit
  of stars: 117, 118
    by annual parallax: *see* annual parallax
    by apparent magnitudes: 6-7, 17, 79-83, 93, 100
    by comparison with Sun: 2, 5, 6, 8, 34, 71, 75-77, 80, 84, 85, 90 (Gregory); 5-6, 8 (Huygens)
    by photometry: 7, 93 (W. Herschel); 7 (J. Herschel)

  of borders of Galaxy: 2, 3, 15-16, 114, 131
  of centre of Galaxy (Shapley): 157-9, 183, 184
  of remote clusters: 3, 154 (W. Herschel)
  of M13: 181
  of Andromeda Nebula (M31): 161, 164, 168
Dollond, J.: 38, 39
γ Draconis: 7-8, 29-33, 34
Duncan, J. C.: 169

Earth, interior of (Halley, Whiston, Wright): 112
Eddington, A. S.: 182, 184
Edmondson, M. R.: 180
Elliott, C. A.: 180
Encke, J. F.: 15
Eyepieces, magnifications of W. Herschel's: 127, 137

Fabricius, D.: 23, 24
Fechner's Law: 85
Ferguson, J.: 57
Fire, role of in the Universe (Wright): 68
Flamsteed, J.
  and annual parallax: 7, 8
  observation of Algol: 44
  star catalogue: 11, 33, 50, 51, 130
Fleming, W. P.: 168, 169
Fraunhofer, J.: 10, 11
Frost, E. B.: 178
Fullenius, B.: 24

Galaxies: 6, 104, 125, 139, 154, 168, 175, 178, 182-4, 185, 186; *see also* the Galaxy
Galaxy, the: 15, 16-17, 68, 84-85, 86, 99, 101, 103-9, 111-12, 113-14, 115, 117-21, 122, 133, 154, 175, 182-5
  "fund of new stars": 23
  globular clusters define centre (Shapley): 157-9, 183, 184
  outline of: 2, 3, 15-16, 114, 131-2
  see also Milky Way; galaxies
Galileo: 9, 15, 17
α Geminorum: 61
β Geminorum: 56
γ Geminorum: 59, 60, 65
George III, King: 129, 137
Gill, D.: 12
God, role of
  according to Derham and Whiston: 3, 67-68, 101-2
  according to Lambert: 2
  according to Newton: 2, 67, 71, 72-74, 86, 87-88, 94 n39
  according to Wright: 101-7, 109-13, 119
Goodricke, J.: 13, 38-48, 51
Goodricke, Sir J.: 38
Graham, G.: 31, 33

Index

Gravity
  cause of proper motions: 56, 102
  effect on initial chaos: 68-70 (Kant); 72-73 (Newton, Bentley)
  existence outside solar system proved: 4, 14-15
  and possible collapse of solar system: 87-88, 96, 113
  and possible collapse of star systems: 2, 3, 56, 71, 73-74, 76-78, 86, 87-88, 94 n39, 95-96, 102, 141
  see universe of stars, development with time; cosmogony
"Great Debate" (1920): 157, 175-87
Gregory, D.: 71, 83, 85-86, 94, 95, 98, 99
Gregory, J.
  method of stellar distances: 2, 5, 6, 8, 71, 75-77, 80, 84, 85, 90

Hale, G. E.: 175, 176, 177, 178, 180
Hale, W. E.: 175
Hall, A. R. and M. B.: 84
Halley, E.
  list of nebulae: 125, 127
  list of novae and variables: 37
  and Olbers's Paradox: 83, 86, 93-94, 95-100
  and proper motions: 2, 9, 13, 102
  theory of geology: 112
  theory of nebulae: 125
  theory of stellar magnitudes: 82
Henderson, T.: 11
α Herculis: 50
λ Herculis: 14, 57, 61, 62, 63, 64, 65
Herschel, C.: 129, 130
Herschel, J.:
  asks Pigott to join RAS: 51
  catalogue of nebulae: 143, 145
  drawings of M42: 150
  and Magellanic Clouds: 149
  and Milky Way: 16, 155
  and orbits of binary stars: 15
  and photometry: 7
  his sweeps for nebulae: 137, 143
  his theory of nebulae: 140, 145
Herschel, W.
  Algol, observations of: 43, 44
  annual parallax by double star method: 9, 14
  astronomical unit, value for: 114
  and binary stars: 6, 15
  and black holes: 154, 166
  catalogues of comparative brightnesses of stars: 12, 19 n52, 49
  catalogues of nebulae and star clusters: 134, 137
  changes in M42 observed: 1, 126-9, 130, 133, 134, 136 n50, 137, 154, 157
  comets, formation of: 140-1
  cosmogony: 122, 131, 133, 134, 135, 139, 141, 142
  distribution of nebulae uneven: 156
  and double stars: 6, 9, 14, 15
  galaxies, existence of: 139, 154, 168
  Galaxy "fathomless": 154
  Galaxy, outline of: 2, 3, 15-16, 114, 131-2
  globular clusters, formation of: 134
  knowledge of Wright, Kant, Lambert: 15, 114-15, 121-3
  "laboratories of the universe": 134
  magnifications of eye-pieces: 127, 137
  millions of light-years, clusters at: 3, 154
  Moon considered inhabited: 137
  Messier catalogue of nebulae received: 125-6, 127, 128, 129, 134
  as natural historian: 3, 139, 152
  nebulae and clusters observed: 125-35, 137; see also changes in M42 observed; NGC 1514 observed
  nebulae equated with clusters: 133-4, 135, 154
  nebulae unevenly distributed: 156
  NGC 1514 observed: 134, 139
  observes stars in remote past: 3, 154
  his origins: 143
  photometry by pairs of matching telescopes: 7, 93
  and planetary nebulae: 129-30, 133-4, 136, 139, 154
  planets, formation of: 141
  proper motions imply solar motion: 96; see solar apex determined
  reflector, "small" 20 ft: 38, 125, 126, 129, 137, 138
  reflector, "large" 20 ft: 67, 131, 134-5, 137, 140, 144, 154
  reflector, 40 ft: 16, 67, 137, 139, 141, 154
  reflector, 25 ft: 143, 146
  refractor, 3½ ft achromatic: 130
  relations with Aubert: 63
  relations with Banks: 43, 44
  relations with Goodricke and Pigott: 38, 39, 43, 47, 51
  relations with Maskelyne: 17, 62-63, 137
  relations with Watson: 139; see Messier catalogue of nebulae received
  repulsive forces: 134, 139
  royal patronage: 127
  solar apex determined: 14, 57-66, 123
  solar velocity determined: 14
  speculations, importance of: 153
  star "gages": see statistics of galactic stars
  stars assumed regularly distributed: 2, 15, 16
  stars assumed equally luminous: 6, 16
  stars habitable: 49, 141
  stars, variable, explained: 44, 47, 49-50, 51
  statistics of brightest stars: 3, 96
  statistics of galactic stars: 2, 3, 16, 17, 114, 132
  stellar magnitudes equated with relative distances: 6-7, 79, 82, 93 n38, 100

and stellar structure: 49-50
sweeps for nebulae: 130, 131, 137
as telescope maker: 69, 145
and "true" nebulosity: 130-1, 132, 134, 135, 139, 150
Tycho's nova explained: 134
Uranus, discovery of: 39, 137
variability of α Herculis discovered: 50
visit to Greenwich: 137
Hertzsprung, E.: 182
Hevelius, J.: 13, 24, 25, 29
Hind, J. R.: 151
Hind's Nebula: 151
Hipparchus: 22, 24
Holwarda, J. P.: 24
Hooke, R.
and annual parallax: 5, 7, 8, 29-31, 32
and proper motions of stars and Sun: 13, 66
and stellar magnitudes: 17 n7
Hornsby, T.: 18, 58
Hoyle, F.: 2
Hubble, E. P.: 150, 160-5, 168, 169
Hubble, G.: 173
Huggins, W.: 1, 151-2, 155
Huygens, C.:
his sketch of M42: 127
and stellar distances: 5-6, 8
R Hydrae: 45, 49

Inhabitants
of comets: 112
of Moon: 137
of planets: 141 (surface), 112 (interior)
of stars: 141
of Sun: 49 (surface), 112 (interior)
Intensifier, Shapley's: 179

Jaki, S. L.: 117-18, 119, 121
Jeans, J. H.: 157
Jungius, J.: 24
Jupiter: 3, 5, 33

Kant, I.: 117, 119
his cosmogony: 68-69
and motion of Sun: 66
theory of the Galaxy and hierarchical universe: 15, 69, 96, 102, 110, 111, 112, 114, 120
Kapteyn, J. C.: 16, 176, 181, 182
Keeler, J. E.: 156, 170
Keill, J.: 26
Kepler, J.: 23, 24, 79, 98, 99
Kirch, G.: 38, 39
Kügel, G. S.: 14

"Laboratories of the Universe", Herschel's: 134
Lalande, J.-J. L. de: 44, 134, 136

list of proper motions: 56, 57, 58, 59, 60, 61, 62, 65
Lambert, J. H.: 2, 15, 66, 69, 96, 110, 117-23
Laplace, P.-S., Marquis de: 143, 151, 166
Leibniz, G. W.: 69, 88, 117, 121
Leovitius: 23
Lexell, A. J.: 114
Liapounov, M.: 150
Light
aberration of: *see* aberration of light
does not slow down (Bradley): 35
loss of: 5, 16, 76, 90, 156, 157, 159, 170, 186, 187
nebulae may be formed from starlight: 140
physical nature of (Halley): 93 n39, 97-98, 100
velocity finite: 3, 33; *see* aberration of light
velocity same for direct and reflected (Bradley): 35
Lindenau, B. A. von: 9, 10
Lowell, A. L.: 177, 179
Lucretius: 95
α Lyrae: 8, 65
β Lyrae: 45, 47, 51

M1 (Crab Nebula): 149, 150
M2: 126
M5: 126, 129
M9: 126, 130-1
M10: 126
M11: 128, 129, 181
M12: 126
M13: 126, 128, 181, 182
M14: 126
M15: 126, 129
M16: 126
M17: 131, 132
M19: 126
M22: 126, 181
M24: 126
M26: 129
M27: 130, 132-3
M28: 126
M30: 126, 131
M31: *see* Andromeda Nebula
M33: 159
M37: 126
M42: *see* Orion Nebula
M51: 126, 148, 183
M52: 126, 129
M53: 126
M55: 126
M56: 126, 130
M57: 129
M62: 126
M65: 126
M67: 126
M71: 126
M72: 126

Index

M74: 126
M82: 130
M92: 126
M97: 150
M101: 157, 158, 165
Magellanic Clouds: 149, 185
Maraldi, G. F.: 40, 42, 45
Marius, S. (Mayr): 24, 25
Marsden: 136
Maskelyne, N.
  and Algol: 42, 43, 44
  criticises papers by W. Herschel: 17, 62-63
  list of proper motions: 57, 58, 59, 60, 61, 64, 65
  visited by Goodricke: 45
  visited by W. Herschel: 137
Mathieu, C. L.: 9
Maupertuis, P. L. M. de: 111
Mayer, J. T.: 13-14, 56, 58, 63, 64, 65, 119
Mayr, S.: see Marius
Méchain, P.-F.-A.: 136
Merleau-Ponty, J.: 117, 121
Messier, C.: 125, 126, 128, 129, 130, 134
Michell, J.
  and black holes: 166
  need for catalogues of stellar magnitudes: 19 n52
  probability argument for binary stars: 4, 14-15
  stars with large proper motions may be near: 18 n27
  theory of stellar magnitudes: 100
Milky Way: 15, 16, 84, 103, 108, 111, 112, 113-14, 117, 119, 154, 155, 157, 175; see also the Galaxy
Mira (o Cet): 13, 23, 24-25, 26, 37, 40, 42, 44, 49
Molyneux, W.: 31, 32, 33
Montanari, G.: 25, 40, 42
Moon, inhabitants of: 137

Natural history of the heavens (W. Herschel): 3, 139, 152
Nebulae: 117, 125-74
  changes in: see under Andromeda Nebula; NGC 1555; Orion Nebula
  distribution of: 155, 156, 157, 185-6, 187
  edge-on: 156, 157, 170, 186, 187
  elliptical: 111, 125
  gaseous: 1, 125, 151, 154
  Halley's theory of: 125
  J. Herschel's sweeps for: 137, 143
  W. Herschel's sweeps for: 130, 131, 137
  life-cycle of: 139
  observed by Maupertuis: 111
  Newton's theory of: 84-85
  novae in: 156, 157, 160, 168-73, 186
  planetary: 125, 133-4, 136, 139, 151, 154; discovered by W. Herschel: 129-30
  radial velocities of: 157, 160
  rotations of: 157, 158, 159, 161, 164, 165
  spectra of: 1, 151, 185
  spiral: 17, 125, 148, 150, 156, 157-65, 175, 176, 178, 182-4, 185-7
  see also under individual nebulae; catalogues
Nebular evolution (Jeans): 157
Nebular hypothesis (Laplace): 143, 151
Newcomb, S.: 178, 183, 184
Newton, I.: 95, 98, 121
  cosmology, stellar distances, stellar magnitudes, and analyses of stellar statistics: 2, 3, 5, 6, 7, 8, 13, 14, 34, 67, 71-91, 96, 100, 119
NGC 1514: 134, 139
NGC 1555 (Hind's Nebula): 151
NGC 3293: 185
NGC 4321: 170-1, 173
NGC 4527: 170-1, 173
NGC 4755: 185
NGC 5253: see Z Centauri
NGC 5272: 185
NGC 6543: 151
NGC 6946: 169-73
NGC 7009 (Saturn Nebula): 129-30
Nichol, J. P.: 145, 149
Novae: 2, 13, 22-26, 37
  galactic: 169, 172-3, 186
  Kepler's (1604): 23, 24, 26, 37, 40
  in nebulae: see under nebulae; Andromeda Nebula; Z Centauri
  Tycho's (1572): 12, 22, 23, 24, 25, 26, 37, 134
  Wright's theory of: 112, 114
Nutation, discovered by Bradley: 8, 12, 34

Obscuration: see light, loss of
Occultation: see stars, occultation by Moon
Olbers, H. M. W.: 96; see Olbers's Paradox
Olbers's Paradox: 95-97, 118
  and Newton: 80, 81, 83
  and Halley: 83, 86, 93-94, 95-100
Orion Nebula (M42)
  listed by Halley: 125, 127
  observed by W. Herschel: 130, 132, 133
  observed by Rosse: 145
  sketched by J. Herschel: 150
  sketched by W. Herschel: 126, 127, 128, 129
  sketched by Huygens: 127
  status of: 120, 123, 134, 139, 143, 154, 155
  supposed changes in: 1, 117, 120, 126, 127-8, 130, 133, 134, 136 n50, 137, 149, 150, 151, 154, 157

Palmerston, Lord: 136
Pannekoek, A.: 85
Parsons, L., 4th Earl of Rosse: 150
Parsons, W., 3rd Earl of Rosse: 143, 145, 150, 151, 155, 156, 168
β Persei: see Algol

Photography used in astronomy: 12, 156, 157, 158, 159, 160, 175
Photometry of stars: 7 (J. Herschel); 7, 93 (W. Herschel)
  and stellar distances: 2, 5, 6, 8, 34, 71, 75-77, 80, 84, 85, 90 (Gregory); 5-6 (Huygens)
Piazzi, G.: 8, 9, 18
Pickering, E. C.: 19, 176
Pigott, E.: 18, 37-52, 136
Pigott, N.: 37, 38, 48, 51
γ Piscium: 56, 61, 62
Planets
  formation of: 141
  inhabitants of: 112, 141
  see also under individual planets
Pleiades: 14
Plummer, H. C.: 181
Polaris (α UMi): 7, 8
Pole Star: 7, 8
Pollux (β Gem): 56
Pond, J.: 9
Pound, J.: 31
Prévost, P.: 14, 62, 63, 64, 65
Probability argument (Michell): 4, 14-15
Proctor, R. A.: 16, 155
Procyon (α CMi): 8, 15, 38, 56
Proper motions
  of planetary nebulae: 130
  of stars: 2, 9, 10, 11, 12, 13, 14, 18 n27, 38, 56-66, 86, 96, 102, 117, 119; as clue to nearest stars: 2, 9 (Bessel); 11 (Henderson); 18 n27 (Michell, Hornsby, Pigott); 38 (Pigott); 10 (Struve)
Providence: see God
Ptolemy, C.: 22, 24, 93, 183

Quadrants: 37

Radial velocities of nebulae: 157, 160
Ramsden, J.: 8, 44
Relativity: 175, 176
Repulsive forces: 134, 139
Riccioli, G.: 24
Ritchey, G. W.: 157, 169-70, 171, 172, 173, 181
Robinson, T. R.: 143, 145
Römer, O.: 33, 65
Rosse, Earl of: see Parsons
Russell, H. N.: 161, 176, 177, 178, 179, 180, 182

St Clair, J. R.: 180
Saturn: 127
Saturn Nebula: 129-30
Saturn's rings composed of lesser planets (Wright): 109
Savary, F.: 15
Schröter, J. H.: 123
Schubert, F. T.: 137
R Scuti: 49, 50, 51

Seares, F. H.: 161, 183
Seeliger, H. von: 16
Shapley, H.: 157, 158, 161, 171, 175-84
Simplicity: 1
Sirius (α CMa): 5, 6, 8, 15, 34, 38, 56, 61, 62, 63
Sisson, J.: 37, 48
Slipher, V. M.: 156-7, 169
Smith, R.: 127
Solar system
  motion of: 7, 12, 13-14, 15, 56-66, 102, 107, 110, 111, 117, 123
  stability of: see gravity and possible collapse of solar system
Spectroscopy: 1, 151, 154, 156, 157
Star clusters: 2, 4, 118, 120, 125, 142
  globular: 6, 125, 134, 154, 157-9, 181, 182, 183, 184, 185
  open (galactic): 125, 181, 182
  see also nebulae and under individual star clusters and nebulae
Stars
  dark: 13, 15, 41-45, 47, 50
  dark core of (W. Herschel and Pigott): 49-50, 52
  distances: see distances of stars; annual parallax
  double: 2, 4, 6, 10-11, 12, 13, 14, 15 (binaries); 41-45, 47, 51 (eclipsing binaries); 9-11, 12, 14, 92 n24 (optical)
  hollow interior of (Whiston and Wright): 112, 113
  inhabitants of: 141
  luminosity of: 7
  magnitudes equated with relative distances: 6-7, 17 n7, 79, 82, 93 n38, 100
  masses of: 7, 15
  multiple: 4, 14; see stars, double; star clusters
  nearest, criteria for: 10, 11; 14, 16 (brightest); 2, 9, 10, 11, 18 n27, 38 (large proper motion)
  new: see novae
  occultation by Moon: 40
  photometry: see photometry of stars
  proper motions: see proper motions of stars
  variable: 2, 13, 22-26, 37-52, 112, 114; 37 (catalogue by Halley); 48 (catalogue by Pigott); see Cepheid variables and under individual stars
  velocities of: 7
  white dwarfs: 15
  see also star clusters and under individual stars
Statistics
  of brightest stars: 3, 76-84, 85, 86, 89, 93 n36, 95, 96, 98-100, 132
  of galactic stars: 2, 3, 16, 17, 114, 132
Struve, F. G. W.: 10, 11, 16, 137
Struve, O.: 11, 169
Struve, O. W.: 150, 151

Index

Sun: *see* solar system; astronomical unit; inhabitants of Sun; stars, dark core of
Sunspots as explanation of variable stars
   in seventeenth century: 13, 22, 25, 37
   Goodricke and Pigott: 13, 42, 45, 47, 48, 49, 50

α Tauri: *see* Aldebaran
Telescopes
   Crossley reflector: 175
   Fraunhofer equatorial at Dorpat: 10
   Fraunhofer heliometer: 10
   Goodricke's 2½ft achromatic: 39
   Goodricke's equatorial: 44
   W. Herschel's 3½ft achromatic refractor: 130
   W. Herschel's "small" 20ft reflector: 38, 125, 126, 129, 137, 138
   W. Herschel's "large" 20ft reflector: 67, 131, 134-5, 137, 140, 144, 154
   W. Herschel's 40ft reflector: 16, 67, 137, 139, 141, 154
   W. Herschel's 25ft reflector: 143, 146
   Mt Wilson 60-inch reflector: 160, 168, 176, 181
   Mt Wilson 100-inch reflector: 160, 168, 176
   pairs of telescopes used by W. Herschel: 7, 93 n36
   meridien refractor at Dorpat: 10
   Rosse's 3ft reflector: 143, 144
   Rosse's 6ft reflector: 144-5, 147, 151, 168
   zenith telescopes: 29-33
Time
   changes with: 2-3; *see* universe of stars, development with time
   observation of remote past: 3, 154
Transit instruments: 37, 48
Transits of Venus and Mercury: 37, 48
Trigons: 23
Troughton, E.: 9
Tycho Brahe: 13, 22, 23, 24

Uniformity: 2, 6
   of stars: 2, 16, 34, 68; *see* photometry of stars and stellar distances
   of stellar distribution: 2, 15-16, 73, 76-84
Universe of stars
   development with time: 3, 68-70, 72, 122, 131, 133, 139, 141

   hierarchical: 110, 112, 113, 120 (Wright); 118, 120 (Lambert); 15, 69, 96, 102, 110, 111, 112, 114, 120 (Kant)
   number of stars in: 67, 96
   orderly arrangement of: 67-68, 101-2 (Derham and Whiston); 119 (Lambert); 78, 84 (Newton); 104, 115 (Wright); *see also* statistics of brightest stars
   stability of: 2, 13, 67, 68, 71, 72-74, 78, 86, 88, 95, 96, 102, 121, 122
   see Olbers's Paradox
Uranus: 39, 40, 137
α Ursae Minoris: 7, 8
Utenhove, J. M. C. van: 117

Van Maanen, A.: 157-9, 161, 164-5
Vega (α Lyr): 8, 10, 65
β Virginis: 38
Volcanoes in the sky (Wright): 112-14
Voltaire, F. M. A. de: 71

Wagman, N. E.: 180
Walters, M. C.: 180
Watson, W., jr: 126, 127, 128, 136, 139
Wendell, O. C.: 169
Whiston, W.: 68
   and cosmic order: 67, 101-2
   and gravitational collapse: 3
   and inhabited interiors of Sun, planets and comets: 112
   and motion of solar system: 13
   and variable stars: 13
Whiteside, D. T.: 88
Wilson, P.: 121
Wolf, M. F. J. C.: 184
Woolfson, M. M.: 52
Wright, T., of Durham: 55, 69, 117
   his cosmology: 15, 66, 68, 96, 101-15, 119, 120

Ximenes, abbé: 136

Young, C. A.: 149

Zebergs, V.: 169
Zone of avoidance: *see* nebulae, distribution of

Index

Sun, see solar system, astronomical unit
inhabitants of Bonaparte, dark ages of
telescopes, as explanation of variable stars
in astronomy, central, 31, 32, 74, 82
Toaldo-ke and Lagoon, 13, 42, 44, 47, 48, 49, 50

to Tenerife Aldebaran
Telescope
Crossley reflector, 175
Fraunhofer equatorial at Dorpat, 10
Fraunhofer heliometer, 10
Goodricke's 25ft achromatic, 15
Goodricke's equatorial, 44
W. Herschel's 15ft achromatic refractor, 120
W. Herschel's "small" 20ft reflector, 38, 129, 136, 139, 137, 138,
W. Herschel's "large" 20ft reflector, 67, 131, 134, 5, 142, 140, 144, 154
W. Herschel's 40ft reflector, 16, 6, 131, 139, 141, 151
W. Herschel 25ft reflector, 135, 146
M. Wilson 60-inch reflector, 170, 168–171, 181
M. Wilson 100-inch reflector, 170, 169, 170
pairs of telescopes used in, by Herschel, 7, 97, 146
meridian refractor at Dorpat, 10
Rosse's 3ft reflector, 143, 144
Rosse's 6ft reflector, 144–5, 147, 151, 156
zenith telescopes, 29, 31
Time,
changes with, 2, 3, see universe of stars, development with time
observation of remote past, 2, 154
Transit instruments, 7, 44
transits of Venus and Mercury, 3, 48
Tucana, 21
Tousphton, E. B.
Tycho Brahe, 13, 22, 23, 24

Uniformity, 2, 6
of stars, 4, 6, of stars, see photometry of stars, and stellar distances
of stellar distribution, 2, 13–15, 73, 76–84
Universe of stars
development with time, 3, 68–70, 72, 122, 131, 134, 138, 140

Variable stars, 136

Tonge, C. A., 146

Zebelin, V., 169
Zone of avoidance, see nebulae, distribution of

Wright, T., of Durham, 55, 69, 117
in cosmology, 13, 56, 65, 69, 101–15, 119, 120
Wolf, M. F.
Wolf, C., 121
Wilson, P., 121
Whiteside, D. T., 48
and variable stars, 13
and motion of solar system, 13
misidentified in work of Sum, planets and comets, 112
and gravitational collapse, 3
and cosmic order, 97, 101-2
Winston, W. 98
Wendell, O. C., 167
Watson, W. Jr., 126, 127, 128, 136, 139
Walters, M. C., 190
Wachmann, A. E., 189

Yolanets, E. M. A., 64–21
Yerkes proceedings/refractor, 172–14
p Viranier, 36
Vest is p. 10, 65, 68, 137
Van Maanen, A. 137–9, 161, 164–5

Tsonkova, T. M. C., var. 177
a Linier Minoris, 8
Urania, 39, 40, 147
see Olbers's Paradox
abundance of, 2, 13, 67, 68, 71, 72, 74, 78, 86, 88, 95, 96, 102, 121, 122
star sizes of brightest stars
578, 84 Crawi ort, 104, 115(Weichsel, see old (Deutson and Whaleon), 115 (Lamber)
(number of stars in, 67, 70
modern arrangement of, 67–84, 101–2
(122–115, 129(Kant)
120 (Lamberts), 15, 69, 96, 102, 110, 111, bicentennial, 110, 112, 113, 130 (Wright), 118